EXPLORING

SCIENCE AND MEDICAL DISCOVERIES

Space Exploration

Other books in the
Exploring Science and Medical Discoveries series:

Antibiotics
Cloning
Gene Therapy
Vaccines

SCIENCE AND MEDICAL DISCOVERIES

Space Exploration

Nancy Harris, *Book Editor*

Bruce Glassman, *Vice President*
Bonnie Szumski, *Publisher*
Helen Cothran, *Managing Editor*
David M. Haugen, *Series Editor*

GREENHAVEN PRESS
An imprint of Thomson Gale, a part of The Thomson Corporation

Detroit • New York • San Francisco • San Diego • New Haven, Conn.
Waterville, Maine • London • Munich

For more information, contact
Greenhaven Press
27500 Drake Rd.
Farmington Hills, MI 48331-3535
Or you can visit our Internet site at http://www.gale.com

Cover credit: David A. Hardy/Science Photo Library
Cover caption: This illustration depicts the unmanned solar spacecraft *Ulysses*, passing
        through the tail of Comet Hyakutake in 1996.
NASA, 16, 56, 62, 79

LIBRARY OF CONGRESS CATALOGING-IN-PUBLICATION DATA
Space exploration / Nancy Harris, book editor.
p. cm. — (Exploring science and medical discoveries)
Includes bibliographical references and index.
ISBN 0-7377-2836-1 (pbk. : alk. paper) — ISBN 0-7377-2835-3 (lib. : alk. paper)
1. Astronautics—History. 2. Outer space—Exploration—History. I. Harris, Nancy, 1952– . II. Series.
TL787.5.S63 2005
629.4'09—dc22   2004054137

CONTENTS

Foreword   10

Introduction: Is There Value in Space Exploration?   13

Chapter 1: Precursors and Early Space Exploration

1. Precursors to Space Exploration
by William E. Burrows   22
Galileo began to bridge the gap between humans and space with his development of the telescope in the first decade of the seventeenth century. His work prompted other theorists to hypothesize ways of traveling into space.

2. Progenitors of Space Flight
by Ray A. Williamson and Roger D. Launius   28
At the beginning of the twentieth century, three men paved the way for mankind's flight into space: a Russian schoolteacher named Konstantin Tsiolkovsky, German scientist Hermann Oberth, and American rocket scientist Robert H. Goddard.

3. The First Satellite and Animal in Space
by William J. Walter   34
In 1957 the USSR gained the upper hand in the space race with the United States when they launched the first space satellite *Sputnik I*. A month later, the Russians sent a satellite with the first dog into orbit.

4. The Formation of the National Aeronautics and Space Administration
by Stephen J. Garber and Roger D. Launius   41
President Eisenhower formed the National Aeronautics and Space Administration in 1958 in response to the

USSR's achievements in space. NASA garnered financial and technical support for space exploration and put the United States back in the space race.

Chapter 2: Problems and Successes of Putting Humans in Space

1. The First Human in Space
by Peter Bond   46
Russian cosmonaut Yuri Gagarin became a legend in his time by being the first human in space on April 12, 1961. Gagarin's spacecraft made one trip around the Earth and returned in less than two hours.

2. Vowing to Put a Man on the Moon
by John F. Kennedy   54
In a special address to Congress on May 25, 1961, President John F. Kennedy made a now famous speech in which he vowed that the United States would put a man on the moon before the decade was over.

3. The Significance of the Apollo Program
by Carl Sagan   58
A word renowned scientist explains that the Apollo program captured the imagination of the world and enhanced the reputation of the United States by landing a man on the moon.

4. Columbia: The World's First Space Shuttle
by Ed Magnuson   65
In April 1981 the world's first reusable spacecraft, the space shuttle *Columbia*, successfully landed in California's Mohave Desert. The successful shuttle mission was a much-needed boost to America's lagging pride and spirit.

5. The Mir Space Station
by Clay Morgan   73
The Russian's *Mir* space station was in orbit around Earth for fifteen years and served as the first large-

scale technical endeavor between Russia and the
United States.

6. The Space Shuttle Challenger Disaster
by Nick Greene   77
The space shuttle *Challenger* explosion of 1986 was
a life-altering event for many who witnessed it. The
Challenger was unique in that it carried the first lay-
person on board, elementary schoolteacher Christa
McAuliffe.

7. Learning from the Columbia Space Shuttle Accident
by Sean O'Keefe   82
NASA's administrator gives the final report on the *Columbia* space shuttle accident. He emphasizes that
NASA will comply with the recommendations of the
Columbia Accident Investigation Board and that the
space program will continue.

Chapter 3: Unmanned Space Exploration

1. The First Spacecraft to Search for Life on Mars
by U.S. News & World Report   91
An interview with an internationally recognized planetary scientist reveals the significance of the *Viking I* information sent to scientists from Mars.

2. The Voyagers' Deep Space Exploration
by Ed Stone, as told to A.J.S. Rayl   97
The project scientist for the *Voyager 1* and *2* missions
to the outer planets says the two probes sent massive
amounts of information to scientists and made unexpected and exciting discoveries.

3. The Hubble Telescope Has a Questionable Future
by Faye Flam   106
The Hubble telescope has been orbiting Earth since

1990, but its future is questionable because of the
costly servicing missions.

4. The Mars Observer Disappears

by John Travis, Christopher Anderson, and Jon Cohen 111
In 1993 the scientific community was devastated when
they lost contact with the Mars *Observer* only three
days before the spacecraft was to enter the martian
orbit.

5. Rovers Reveal That Mars Was Once Wet

by Marsha Walton 122
Two space rovers landed on Mars in January 2004 and
verified that water had existed on the planet at one
time. This fact suggests that there may have been life
on the red planet.

Chapter 4: Current Controversies and Outlooks in Space Exploration

1. Human Space Explorers Are Superior to Robots

by Robert Zubrin 126
Human presence in space is necessary because robots
are not capable of performing some of the tasks that
further scientific discovery.

2. Human Accomplishments on the International Space Station Will Be Inconsequential

by Robert L. Park 131
Experiments conducted on the International Space Station will be unimportant because some of them have
already been done. Instead of sending humans to the
ISS, the space program should concentrate on using
robots for exploration.

3. NASA's Projects Are Expensive, Wasteful, and Inefficient

by Josh London 137
NASA has become a massive federal bureaucracy that

overspends, has poor programs, and conducts worthless scientific experiments.

4. NASA Has Made Important Technological and Economic Contributions
by Todd Wilkinson   143
NASA has made significant contributions not only to aeronautics and space exploration but to the economic and technological well-being of the general public as well.

5. Space Exploration Should Be Supported by a Space Tourism Business
by Buzz Aldrin   150
Space tourism could easily support the twenty-first century space program.

6. Space Should Be Protected as a Wilderness
by Ryder W. Miller   156
Those involved in space exploration are not conscientious enough about the space environment. NASA and the United Nations should support a program of "astro-environmentalism."

Chronology   161
For Further Research   166
Index   171

Most great science and medical discoveries emerge slowly from the work of generations of scientists. In their laboratories, far removed from the public eye, scientists seek cures for human diseases, explore more efficient methods to feed the world's hungry, and develop technologies to improve quality of life. A scientist, trained in the scientific method, may spend his or her entire career doggedly pursuing a goal such as a cure for cancer or the invention of a new drug. In the pursuit of these goals, most scientists are single-minded, rarely thinking about the moral and ethical issues that might arise once their new ideas come into the public view. Indeed, it could be argued that scientific inquiry requires just that type of objectivity.

Moral and ethical assessments of scientific discoveries are quite often made by the unscientific—the public—sometimes for good, sometimes for ill. When a discovery is unveiled to society, intense scrutiny often ensues. The media report on it, politicians debate how it should be regulated, ethicists analyze its impact on society, authors vilify or glorify it, and the public struggles to determine whether the new development is friend or foe. Even without fully understanding the discovery or its potential impact, the public will often demand that further inquiry be stopped. Despite such negative reactions, however, scientists rarely quit their pursuits; they merely find ways around the roadblocks.

Embryonic stem cell research, for example, illustrates this tension between science and public response. Scientists engage in embryonic stem cell research in an effort to treat diseases such as Parkinson's and diabetes that are the result of cellular dysfunction. Embryonic stem cells can be derived from early-stage embryos, or blastocysts, and coaxed to form any kind of human cell or tissue. These can then be used to replace damaged or diseased tissues in those suffering from intractable diseases. Many researchers believe that the use of embryonic stem cells to treat human diseases promises to be one of the most important advancements in medicine.

However, embryonic stem cell experiments are highly controversial in the public sphere. At the center of the tumult is the fact
that in order to create embryonic stem cell lines, human embryos
must be destroyed. Blastocysts often come from fertilized eggs that
are left over from fertility treatments. Critics argue that since blastocysts have the capacity to grow into human beings, they should
be granted the full range of rights given to all humans, including
the right not to be experimented on. These analysts contend, therefore, that destroying embryos is unethical. This argument received
attention in the highest office of the United States. President
George W. Bush agreed with the critics, and in August 2001 he announced that scientists using federal funds to conduct embryonic
stem cell research would be restricted to using existing cell lines.
He argued that limiting research to existing lines would prevent any
new blastocysts from being destroyed for research.

Scientists have criticized Bush's decision, saying that restrict-
ing research to existing cell lines severely limits the number and
types of experiments that can be conducted. Despite this considerable roadblock, however, scientists quickly set to work trying to
figure out a way to continue their valuable research. Unsurpris-
ingly, as the regulatory environment in the United States becomes
restrictive, advancements occur elsewhere. A good example concerns the latest development in the field. On February 12, 2004,
professor Hwang Yoon-Young of Hanyang University in Seoul,
South Korea, announced that he was the first to clone a human
embryo and then extract embryonic stem cells from it. Hwang's
research means that scientists may no longer need to use blastocysts to perform stem cell research. Scientists around the world
extol the achievement as a major step in treating human diseases.

The debate surrounding embryonic stem cell research illustrates
the moral and ethical pressure that the public brings to bear on the
scientific community. However, while nonexperts often criticize
scientists for not considering the potential negative impact of their
work, ironically the public's reaction against such discoveries can
produce harmful results as well. For example, although the outcry
against embryonic stem cell research in the United States has resulted in fewer embryos being destroyed, those with Parkinson's,
such as actor Michael J. Fox, have argued that prohibiting the development of new stem cell lines ultimately will prevent a timely
cure for the disease that is killing Fox and thousands of others.

Greenhaven Press's Exploring Science and Medical Discoveries series explores the public uproar that often follows the disclosure of scientific advances in fields such as stem cell research. Each anthology traces the history of one major scientific or medical discovery, investigates society's reaction to the breakthrough, and explores potential new applications and avenues of research. Primary sources provide readers with eyewitness accounts of crucial moments in the discovery process, and secondary sources offer historical perspectives on the scientific achievement and society's reaction to it. Volumes also contain useful research tools, including an introductory essay providing important context, and an annotated table of contents enabling students to quickly locate selections of interest. A thorough index helps readers locate content easily, a detailed chronology helps students trace the history of the discovery, and an extensive bibliography guides readers interested in pursuing further research.

Greenhaven Press's Exploring Science and Medical Discoveries series provides readers with inspiring accounts of how generations of scientists made the world's great discoveries possible and investigates the tremendous impact those innovations have had on the world.

Is There Value in Space Exploration?

"The great adventure worthy of the twenty-first century is to explore where no human can ever set foot. In the entire history of humanity, we could never do that before. But with modern technology, we can explore places where no human being can ever go. This is the exciting future we have in space exploration."[1] With these words, physics professor Robert Park expresses the feelings of many who place great value in the exploration of space. Since humankind's beginnings on Earth, people have been pondering the stars and planets. Scientists suggest that humans explore space because of a most basic instinct for survival; that is, the more that humans expand their territory, the less chance for their extinction. Space allows humans a place to expand into. It allows for new resources and knowledge, along with the ability to inhabit other worlds. Humankind is presently living on the single, small, and fragile planet Earth, which scientists say could someday be hit and destroyed by meteors. Humans' expansion into space, to other planets, the moon, or other places in the universe, could determine the ultimate survival of the human species.

Aside from the possible value space exploration offers for human survival, some people find other values in exploring space. These values, like beauty, are in the eye of the beholder. To some, space exploration provides a sense of pride, a stirring of creative spirit and imagination, a vicarious experience of wandering and wondering, and a fulfillment of the basic human urge to explore the unknown. To others, space exploration holds intrinsic scientific or technological merit. To still others, it is a means of economic gain or of attracting young people to the field of science and engineering. Space exploration may connote competition or cooperation. For many, it is a combination of several or all of these. And for some people, there is very little or even no value in space exploration.

The inspirational value of space exploration was captured in the words of the visionary space-age author Arthur C. Clarke when he said, "Where there is no vision, the people perish. . . . Men need the mystery and romance of new horizons almost as badly as they need food and shelter."[2] Some people believe that this is the most important aspect and contribution of space exploration. However, scientists find many practical reasons for exploring space as well. Scientists say that exploring other worlds helps them better understand Earth. "Having a better understanding of other worlds, other planets and their satellites helps us understand the history and possible future of our own planet. 'Why is Mars such a cold place and Venus such a hot place?' These are conditions that may affect the climate here on earth. These are important issues that affect us,"[3] says assistant professor of chemistry Dr. William Hagan. Seeing Earth from a different perspective, such as in the photos of Earth taken from the moon and from the outer solar system, gives new insight into this planet's place in the universe. Orbital viewing devices such as the Edwin Hubble Telescope and the future James Webb Telescope, built to take Hubble's place, allow scientists to see back to the beginnings of the universe, thus deepening the understanding of the universe itself.

Technical and Scientific Value

Practical reasons for exploring space include many technical and scientific aspects. The most immediate technical and scientific contributions are from advances in the space field itself. Astronomy, rocket science, and astrophysics are all fields that have contributed to and expanded because of space exploration. Scientific developments in these areas include newer and more advanced manned and unmanned spacecraft, space probes and rovers, telescopes, observatories, and satellites.

Other technical developments are peripheral to space exploration yet some believe just as significant. These are called spin-off technologies—technologies that have developed as a result of research and testing done by the National Aeronautics and Space Administration (NASA) in the space programs over the last several decades. These technologies have produced million-dollar industries that have contributed to the economy not only by manufacturing new products but by creating jobs as well. According to NASA sources, since 1976, between forty and fifty commercial products created by spin-off industries contribute annually to developments in health and medicine, industry, consumer goods, computer technology, and the environment. Consumer goods such as cordless power tools and appliances, freeze-dried foods, and smoke detectors, are some of the most successful commercial spin-offs of space-based technology. Virtual reality is another by-product of research in space exploration.

Some of NASA's most impressive spin-offs include communication satellites, spy satellites, weather satellites, and satellite navigation systems called global positioning networks. Among the many uses for satellites, technicians have used NASA's satellite remote sensing technology to benefit the environment by locating and mapping forest fires. In the field of medicine, space research has made substantial contributions. For example, from the technology NASA developed to computer-enhance pictures of the moon for the Apollo program came computerized axial tomography (CAT) scanners, used to create three-dimensional images of the inner human body, as well as magnetic resonance imaging (MRI) technology, which also allows doctors to see into the human body in new ways. Technology used in space shuttle fuel pumps led to the development of miniaturized artificial heart pumps. Also part of the space shuttle program was special lighting technology developed for plant growth experiments that is now used to treat brain tumors in children.

International Competition and Cooperation

Technological achievements in space may have received their first strong impetus in the early days of the space race after World War II, when competition between the United States and Russia, engaged in the Cold War, spilled over into a race to conquer space. In the 1950s, scientists agreed on an international effort to study Earth from space with satellites, spawning a fierce competition between the United States and Russia to be the first to produce the satellites. It was the Russians who put not only the first artificial satellite into orbit, but also the first human being into space. Americans saw these Russian accomplishments as major blows to

America's national pride and prowess. The competition heightened between the two countries with the creation of NASA and the U.S. determination, through the Apollo program, to put the first human on the moon. At this time, in the 1960s and 1970s, the U.S. government and the American public wholeheartedly supported the space program. In what some call the golden age of space exploration, seemingly boundless amounts of money, time, and energy were devoted to the U.S. efforts in the race to space. It was a time of boldness enhanced by a sense of adventure that birthed what some say are the most remarkable achievements of the twentieth century. During this time, the first American orbited Earth, the United States made its historical man-on-the-moon landing, scientists launched Telstar, the first telecommunications

In July 1969 Neil Armstrong took this photograph of fellow astronaut Edwin "Buzz" Aldrin as they became the first humans to walk on the moon.

satellite, and the space shuttle program was inaugurated.

Since that time, attitudes toward space exploration have changed. The Cold War has ended, and it is now cost rather than competition or scientific achievement that determines whether a nation will adopt a major space program. Although competition still exists among nations exploring space, the value of international cooperation is significant. Space exploration nourishes a positive international cooperative spirit. Possibly the best example of this international cooperation is the work that has been done among nations on the International Space Station (ISS), the largest cooperative scientific program in history. The nations involved in the ISS program must work on a scientific and technical level that is unprecedented. The complex project involves hundreds of contractors and subcontractors around the world. Although the United States and Russia play the largest roles in ISS construction and operations, fourteen other nations are also involved in the project. Eleven of these are European nations. According to Diter Isakeit, the European Space Agency director of manned space flight, "In fifty years, we have turned our former enemies into friends. We freely cooperate on many projects in Europe. We are on the leading edge of global cooperation. Now, we introduce that to the rest of the world."[4]

Working together on space projects, the United States and Russia, formerly staunch rivals, have collaborated on projects such as the 1995 orbital docking of the U.S. spacecraft *Discovery* with the Russian space station *Mir*. The occasion elicited a magnanimous comment from *Mir*'s commander, Alexander Viktorenko: "We are all one. We are human."[5] Indeed, the *Mir* (meaning "peace") space station was host to more than one hundred people from more than a dozen nations over the course of its fifteen-year lifetime.

Many of those working in the field of space exploration express an allegiance not only to their nation but to all of humanity as well. When Neil Armstrong was asked how he felt while standing on the moon saluting the American flag after taking the first historic step there in 1969, he replied, "I suppose you're thinking about pride and patriotism. But we didn't have a strong nationalistic feeling at that time. We felt more that it was a venture of all mankind."[6] The United Nations (UN) also supports this attitude. The text of a UN declaration of international cooperation in outer space states that the UN supports "the principle that the exploration and use of outer space, including the moon and other celestial bodies, shall be carried out for the benefit and in the interests of all countries, irrespective of their degree of economic or scientific development, and shall be the province of all mankind."[7]

Objections and Controversy over Space Exploration

Although many people see great value in space exploration, others argue that the negative economic, ethical, social, and practical consequences of exploring other planets outweigh any potential benefits. One of the biggest objections is that sending humans into space is too expensive and, compared to using robots, dangerous as well. Critics argue that it is the robots that fly the spacecraft and that humans are merely expensive passengers. According to these critics, humans are handicapped by the space suits they must wear, making the humans inferior to robots as explorers. Critics have numerous other objections to the expense of NASA projects. For example, the U.S. space shuttle program was years behind schedule, as is the ISS project. Both went millions of dollars over budget, a phenomenon that dampens public enthusiasm in an area where public interest is already lagging. A 2004 Mars mission costing $266 million produced pictures that were unexciting to the average American. Surveys show that the majority of Americans are unenthusiastic about going to Mars or back to the moon. Even if they do support these projects, they feel that the cost is prohibitive.

Take Care of the Earth First

In addition to the financial objections, some scientists and other Americans believe that resources should be used for the improvement of planet Earth rather than on space exploration. They believe that because humankind has so spoiled the home planet, spending on space exploration is unjustified and even immoral. These critics of space exploration believe that people have a moral obligation to eradicate poverty, racism, and social and economic problems before exploring space. Some scientists in fields outside of space exploration not only agree with this viewpoint, but cite an additional problem of misplaced priorities in the research done in the field of space exploration as compared to research done on Earth itself. For example, scientists have better maps of the surface of Mars than of Earth's ocean floor, and investigations of

Earth and its oceans have been sparse when compared to investigation for water elsewhere in the universe. To some scientists, this situation is unacceptable.

In addition, people have raised ethical questions about what space researchers are planning to do to the environments of other planets. For example, space scientists are doing research on terraforming, an effort to create atmospheres around the planets Mars and Venus in order to make them habitable for humans. However, scientists have not thoroughly explored these planets for indigenous life, raising ethical questions about damaging the environments and microbial life of the planets. Some people even wonder how humans can be expected to maintain atmospheres around other planets when Earth's atmosphere itself is so poorly maintained.

Critics also point out another problem—space debris in near Earth orbit. This man-made space junk—remnants of old spaceships, satellites, and rockets—weighs millions of pounds. It is currently orbiting Earth at about twenty thousand miles per hour, at altitudes from hundreds of miles to many thousands of miles above the planet's surface. According to environmentalist Russell D. Hoffman, "The problem of space debris is a shameful relic of several hundred explosions and mishaps in outer space. Getting past it is the deadly gamble of every space mission. There is no defense against it."[8]

These many objections to the possible value of exploring space represent common differences that may occur in any field of human endeavor. While the outcome is uncertain, some believe it is the values held by the younger generation that will determine the future of space exploration. In the words of Deputy Assistant to the President for Economic Policy W. Bowman Cutter, at a session on the economic value of space exploration, "I think the ultimate answer to the question that all of us are considering—what is the value of exploration—winds up being a question that you cannot answer. It winds up being a question about a new kind of relevance, but more and more, it's a problem that you work on every day. It's not one you answer."[9]

Notes

1. Robert L. Park, "The Virtual Astronaut," *New Atlantis*, Winter 2004, p. 90.
2. Quoted in William J. Walter, *Space Age*. New York: Random House, 1992, p. xii.

3. Quoted in Kate Blain, "See Value in Space Exploration," *The Evangelist*, April 2003, www.evangelist.org/archive/hm/0805moon.htm.

4. Quoted in Stanley Holmes, "International Space Station," *Aerospace*, March 1999, p. 16.

5. Quoted in Mark Carreau, "Mir, Shuttle Have Close Encounter," *Houston Chronicle*, February 7, 1995.

6. Quoted in Tad Daley, "Our Mission on Mars," *Futurist*, September/October 2003, p. 29.

7. United Nations General Assembly, GA Resolution 51/122: "Declaration on International Cooperation in the Exploration and Use of Outer Space for the Benefit and in the Interest of All States, Taking into Particular Account the Needs of Developing Countries," March 12, 1996, UN Office for Outer Space Affairs, www.oosa.unvienna.org/SpaceLaw/gares/pdf/ARES_51_122E.pdf.

8. Russell D. Hoffman, "Space Debris," www.animatedsoftware.com/spacedeb.

9. Molly Macauley, W. Bowman Cutter, and Daniel F. Burton Jr., "Session 2: What Is the Economic Value of Space Exploration? (Part 1)," Center for Mars Exploration, http://cmex-www.arc.nasa.gov/CMEX/data/vse/session2.html.

CHAPTER 1

Precursors and Early Space Exploration

Precursors to Space Exploration

By William E. Burrows

In the following selection, William E. Burrows says that Galileo bridged the gap between humans and space when the seventeenth-century Italian scientist and inventor created the telescope. With his telescope, Galileo discovered Jupiter's four largest moons, demonstrating that everything in the solar system did not orbit the Earth. In his historic pamphlet *The Starry Messenger* published in 1610, Galileo also reported that the Moon's surface was rough and full of cavities, something like the Earth's, suggesting that people might one day be able to visit there.

Burrows explains that for the next several centuries, science fiction writers such as H.G. Wells and Jules Verne stimulated the public's imagination with fantastic space travel stories. These stories along with continuing scientific discoveries prompted the rocketry work of scientists in the late nineteenth century.

William E. Burrows is a professor of journalism at New York University and the founder and director of the university's Science and Environmental Reporting Program. Burrows has written about aviation and space for the *New York Times*, the *Washington Post*, and the *Wall Street Journal.* He has been writing for more than twenty years and is contributing editor for *Air & Space/Smithsonian.* Burrows has authored seven books including *Deep Black*, an award-winning work on spying from space.

The urge to leave Earth to explore other worlds is so old that it has no discernible beginning. But the will to explore, and the means of doing so, were unreconciled until relatively recently. And the first machine to bridge the gap was not one that defied gravity but one that defied distance: the telescope.

In the first decade of the seventeenth century, Galileo turned his homemade telescope to the night sky and beheld the new worlds there from a vantage point closer than any who had come before. In the words of the physicist Philip Morrison, he thereby finally "healed the ancient split between the heavens and the Earth."

Then Galileo was driven to get closer, and closer still. Having raised the magnification of the three-power telescope that had shortly before been invented in Holland to nine-power by August 1609, the professor of mathematics and his craftsman-assistant pushed the cutting edge of the technology still further: to twenty-power by November and then to thirty-power by January 1610. That year he discovered the four largest of Jupiter's moons, and, in doing so, showed that not everything in the solar system orbited Earth. He also discovered that the Sun had spots—"blemishes"— which contradicted the religious and philosophical view, popularized by Aristotle, that it was perfect. These observations, in addition to others showing that Venus went through phases, like Earth's own moon, were another strong indication that Earth was not the center of the universe, as religious dogma had it, but that it circled the Sun just as the Polish astronomer Copernicus had insisted as far back as 1543.

While the origin of astronomy is lost in time, the use of optics to explore the heavens dates from Galileo. Metaphorically speaking, he and his telescope were the forerunners of today's rocket-propelled space probes. His observations of planetary motion, meticulously recorded on grids, were the equivalent of the mechanical sensors—the eyes and ears—on a deep space probe. The telescope amounted to being his booster. It was the means by which Galileo's intellect was able to reach the target: it got his "sensors" to where they needed to be. From his time to ours, astronomers have depended upon a succession of improved telescopes to get them ever closer to ever more distant destinations.

Within thirty-five years of Galileo's observations of the Moon, richly detailed maps had been made of the lunar surface showing some three hundred features, including prominent craters. During that time, William Gascoigne mounted a micrometer on his telescope so he could measure the diameters of planets and stars in a way that he called "strangely precise." The English astronomer reported that he got the idea for the measuring device after God— the "All Disposer"—had caused a spider to spin a thread in an open case while he was experimenting with a view of the Sun.

An Explosion of Science

There was an explosion of science from England to Italy in the waning years of the seventeenth century, much of it focused on the heavens. In 1676 Edmund Halley cataloged 341 stars in the Southern Hemisphere. The following year he made the first complete observation of the transit of Mercury. He also calculated the orbit of the comet that bears his name. This was the time of Isaac Newton, the brightest star in the European scientific firmament, and an individual who was to lay the groundwork for the eventual physical move off the planet and into space.

But touching—going there—was the thing. The great voyages of exploration in which Columbus, Vespucci, Balboa, Magellan, Cabot, Drake, and the others prowled their puzzling world, in effect fitting its disparate pieces into a comprehensible whole that could be studied at first hand rather than merely envisioned as abstract geometry, would before long trace a route that came to point upward.

If the New World was similar to the Old in fundamental respects, might not some of the spheres of the night also share characteristics with the place from which they were observed? It was an invigorating idea and one upon which Galileo himself mused. His discovery that the Moon's surface was "uneven, rough, and full of cavities and prominences, being not unlike the face of the Earth" was published in a historic pamphlet, *The Starry Messenger*, in 1610. The implication of Galileo's observations was clear. If the Moon was similar to Earth, then sending people to it was possible, at least theoretically. It would only be a matter of getting there, of finding a way to break free of Earth and embark on an extended voyage.

Going to Space

The difference between the watchers and the wanderers was that the former had to content themselves with observation—with being able to look but not touch—since they had to stay home. But the wanderers had the means to go where they pleased and touch where they went because they had an infinite source of propulsion. They had sailing ships for boosters and they had the power of the wind.

Exploration would be spiritually uplifting, to be sure. But where science was concerned, the difference between merely seeing and being able to touch would also be of fundamental importance. Going to space, instead of merely looking at it, would open the way to a higher order of science because it would allow actual experimentation, rather than only observation.

Problem-solving through observation requires the sifting of vast quantities of data over years or centuries, a complex, protracted, and arduous process, Lloyd V. Berkner and Hugh Odishaw would explain in a pioneering work about space science soon after the space age began. Berkner, an eminent geophysicist, was chairman of the National Academy of Sciences' Space Science Board when he and Odishaw edited the book *Science in Space*, which would appear in 1961. But experimentation "permits the gifted experimenter to devise his measurements in such a fashion as to separate the variable of interest from the many unwanted ones." It is the difference between observing a hundred criminal suspects over a long period of time to discover which exhibits suspicious behavior, a laborious sorting-out process, versus being able to associate with them and devise a specific trap to ensnare the guilty party.

Flights of Fantasy

The longing to touch space, but not be able to do so for lack of a means to get there, caused a frustration that led to flights of fantasy. As early as the second century A.D. the Greek satirist Loukianos, generally known as Lucian of Samosata, wrote an apparent parody of Homer's *Odyssey* called *True History*. The hero, lifted by a giant whirlwind, goes to the Moon and encounters fantastical humans and animals. In 1638 Bishop Francis Godwin's *The Man in the Moone, or a Discourse of a Voyager thither by Domingo Gonzales, the Speedy Messenger* had its hero breaking the bonds of Earth by using trained swans. The story was imitated in the mid-seventeenth century by Savinien Cyrano de Bergerac in two stories that usually appear together as *Voyages to the Moon and the Sun*. Jonathan Swift in turn borrowed from either Godwin or de Bergerac when he penned Gulliver's voyage to Laputa. A spate of science fiction having to do with space travel came out of France in 1865, including Alexander Dumas's *Journey to the Moon* and Achille Eyraud's *Voyage to Venus*. The latter described a propulsion system consisting of a reaction motor that used water under pressure. It wouldn't have moved a real spacecraft an inch, but the essential principle was on the mark. *The First Men on the Moon*, which was written by H.G. Wells and published two years before the Wright Brothers' airplane flew at Kitty Hawk, used an antigravity device to escape the pull of the Earth. And undoubtedly the work that did most to popularize the concept of spaceflight was Jules Verne's *From the Earth to the Moon*, in which a nine-hundred-foot-long cannon sunk in the ground in Tampa, Florida, shot a 19,250-pound manned aluminum shell all the way to the Moon.

Kepler's Abundant Imagination

In many ways, one of the most remarkable of the fictional odysseys in space was written by an astronomer himself, the eminent Johannes Kepler, and was published posthumously by his son in 1634. *Somnium* (Dream), an allegory, was loaded with footnotes that went into specifics about the scientific aspects of a voyage to the Moon. Kepler, himself a pioneering student of optics, was imperial mathematician to Rudolph II in Prague when he discovered the law of elliptical orbits and made important contributions to the understanding of gravity and the tides, which he theorized were created by lunar attraction. His work on gravity, in fact, foretold Newton's law of universal gravitation. Kepler was among the first scientists whose abundant imagination and creative urge spilled over into fiction because physical science alone could not provide an adequate milieu for them.

Galileo sent a copy of his *The Starry Messenger* over the Alps to Kepler soon after it was published. On April 19, 1610, Kepler replied in an open letter titled *Conversation with the Starry Messenger.* "Who would have believed that a huge ocean could be crossed more peacefully and safely than the narrow expanse of the Adriatic, the Baltic Sea or the English Channel?" Kepler asked Galileo in evident reference to the voyages of Columbus and the others that were being made across the Atlantic with increasing frequency. Then Johannes Kepler, grasping the fact that the exploration of the Moon was only an extension of the exploration of Earth, went on to make a prediction that was as prescient as it was hauntingly beautiful. "Provide ship or sails adapted to the heavenly breezes, and there will be some who will not fear even that void [of space]. . . . So, for those who will come shortly to attempt this journey, let us establish the astronomy: Galileo, you of Jupiter,

I of the Moon." As Kepler correctly pointed out, the problem with going to the Moon was not the destination itself, but the means of getting to it.

The Father of Space Travel

The pendulum finally swung back from fantasy to reality in Russia in the last two decades of the nineteenth century when a poor and unknown schoolteacher named Konstantin Eduardovich Tsiolkovsky developed theoretical models for rocket propulsion. He based his work on Newton's calculations, first published in his *Philosophiae Naturalis Principia Mathematica* in 1687. The treatise included a thorough account of the principles of sending objects into orbits around Earth and even included a still-famous diagram illustrating the technique. Tsiolkovsky was trained in physical science and mathematics and had a special interest in astronomy. These interests combined to produce a unique synergy: relatively early on he became an apostle of space travel and spent his life working on the means to achieve it. When he died in 1935, at age seventy-eight, the world outside the Soviet Union was largely ignorant of his many innovations. But he has subsequently been recognized throughout the world as the master theoretician of the space age: the "father of space travel."

Progenitors of Space Flight

By Ray A. Williamson and Roger D. Launius

The efforts of rocket scientists contributed immensely to the field of space exploration. Ray A. Williamson and Roger D. Launius report in this selection that the pioneering efforts of three men paved the way for humankind's flight into space. The earliest was Russian theoretician Konstantin Eduardovich Tsiolkovsky who laid out many of the principles of modern spaceflight in an article published in 1903. Although Tsiolkovsky never experimented with rockets, his theoretical work was the foundation of the Soviet space program and was influential to rocket scientists' later work. In the 1920s, German scientist Hermann Oberth, known as the father of German rocketry, not only published a classic study that discussed almost every phase of rocket travel, but designed a rocket that he believed could reach the upper atmosphere. Also in the 1920s, American scientist Robert H. Goddard pioneered the use of rockets for spaceflight. Skeptics criticized the already secretive Goddard, but the scientist continued building and launching rockets as well as working on rocket propulsion and multistage rockets and eventually registered 214 patents on various rocket components. Goddard's work heralded in the modern age of rocketry and led to the development of later rocket engines.

Ray A. Williamson is a senior research scientist at the Space Policy Institute in Washington, D.C. As a faculty member of the International Space University and the editorial board of *Space Policy,* Williamson has written extensively about the U.S. space program. Roger D. Launius is curator in the Division of Space History at the Smithsonian Institution, Washington, D.C., and is NASA's chief historian.

C uriosity about the universe and other worlds has been one of the few constants in the history of humankind. Prior to the twentieth century, however, there was little opportunity to explore the universe except in fiction and through astronomical observations. These early explorations led to the compilation of a body of knowledge that inspired and in some respects informed the efforts of certain scientists and engineers who began to think about applying rocket technology to the challenge of space flight in the early part of the twentieth century. These individuals were essentially the first space-flight pioneers, translating centuries of dreams into a reality that matched in some measure the expectations of the public that watched and the governments that supported their efforts. During the period between 1926, when Robert H. Goddard launched his first rocket, and 1957, when the first orbital spacecraft was launched, a dedicated group of rocketeers made the space age a reality, although by the beginning of World War II much of the technology was being developed by government organizations as potential weapons.

Progenitors of the Space Age

While the technology of rocketry was moving forward on other fronts, some individuals began to see its use for space travel. There were three great pioneering figures in this category—collectively, they were the progenitors of the modern space age. The earliest was the Russian theoretician Konstantin Eduardovich Tsiolkovsky. An obscure schoolteacher in a remote part of tsarist Russia in 1898, he submitted for publication to the Russian journal *Nauchnoye Obozreniye (Science Review)* a work based upon years of calculations that laid out many of the principles of modern space flight. His article was not published until 1903, but it opened the door to future writings on the subject. In it Tsiolkovsky described in depth the use of rockets for launching orbital space ships. Tsiolkovsky continued to theorize on the subject of space flight until his death, describing in great detail both methods of flight and the technical requirements of space stations. Significantly, he never had the resources—nor perhaps the inclination—to experiment with rockets himself. His theoretical work, however, influenced later rocketeers both in his native land and abroad and served as the foundation of the Soviet space program.

A second rocketry pioneer was Hermann Oberth, by birth a

Transylvanian but by nationality a German. Oberth began studying the nature of space flight about the time of World War I and
published his classic study *Die Rakete zu den Planetenraümen
(Rockets in Planetary Space)* in 1923. It was a thorough discussion of almost every phase of rocket travel. He posited that a
rocket could travel in the void of space and that it could move
faster than the velocity of its own exhaust gases. He noted that
with the proper velocity a rocket could launch a payload into orbit around the earth, and to accomplish this goal he reviewed several propellant mixtures to increase speed. He also designed a
rocket that he believed had the capability to reach the upper atmosphere by using a combination of alcohol and hydrogen as fuel.
Oberth also discussed other aspects of rocketry and the prospects
of space travel. He became the father of German rocketry. Among
his protégés was Wernher von Braun, the senior member of the
rocket team that built NASA's Saturn launch vehicle for the trip
to the Moon in the 1960s.

Using Rockets for Space Flight

Finally, the American Robert H. Goddard pioneered the use of
rockets for space flight. Motivated by reading science fiction as a
boy, Goddard became excited by the possibility of exploring
space. In 1901 he wrote "The Navigation of Space," a short paper
that argued that movement could take place by firing several cannon, "arranged like a 'nest' of beakers" in a single direction. He
tried unsuccessfully to publish this article in *Popular Science
News*. At his high school oration in 1904 he summarized his future life's work: "It is difficult to say what is impossible, for the
dream of yesterday is the hope of today and the reality of tomorrow." In 1907 he wrote another paper on the possibility of using
radioactive materials to propel a rocket through interplanetary
space. He sent this article to several magazines, and all rejected it.
Still not dissuaded, as a young physics graduate student he worked
on rocket propulsion and actually received two patents in 1914.
One was the first for a rocket using solid and liquid propellants
and the other for a multistage rocket.

After a stint with the military in World War I, where he worked
on solid rocket technology for use in combat, Goddard became a
professor of physics at Clark College (later University) in Worces-
ter, Massachusetts. There he turned his attention to liquid rocket

propulsion, theorizing that liquid oxygen and liquid hydrogen were the best propellants but learning that oxygen and gasoline were less volatile and therefore more practical. To support his investigations, Goddard applied to the Smithsonian Institution for assistance in 1916 and received a five-thousand-dollar grant from its Hodgkins Fund. His research was ultimately published by the Smithsonian as the classic study *A Method of Reaching Extreme Altitudes* in 1919. In it Goddard argued from a firm theoretical base that rockets could be used to explore the upper atmosphere. Moreover, he suggested that with a velocity of 17,500 miles per hour (mph), without air resistance, an object could escape Earth's gravity and head into infinity, or to other celestial bodies. This became known as Earth's "escape velocity."

Goddard's Ideas Are Ridiculed

It also became a great joke for those who believed space flight either impossible or impractical. Some ridiculed Goddard's ideas in the popular press, much to the consternation of the already-shy Goddard. Soon after the appearance of his publication, he commented that he had been "interviewed a number of times, and on each occasion have been as uncommunicative as possible." The *New York Times* was especially harsh in its criticisms, referring to him as a dreamer whose ideas had no scientific validity. It also compared his theories to those advanced by novelist Jules Verne, indicating that such musing is "pardonable enough in him as a romancer, but its like is not so easily explained when made by a savant who isn't writing a novel of adventure." The *Times* questioned both Goddard's credentials as a scientist and the Smithsonian's rationale for funding his research and publishing his results.

The negative press Goddard received prompted him to be even more secretive and reclusive. It did not, however, stop his work, and he eventually registered 214 patents on various components of rockets. He concentrated on the design of a liquid-fueled rocket, the first such development, and the attendant fuel pumps, motors, and control components. On 16 March 1926 near Auburn, Massachusetts, Goddard launched his first successful rocket, a liquid oxygen and gasoline vehicle that rose 184 feet in 2.5 seconds. This event heralded the modern age of rocketry. He continued to experiment with rockets and fuels for the next several years. A spectacular launch took place on 17 July 1929, when he flew the first instrumented payload—an aneroid barometer, a thermometer, and a camera—to record the readings. The launch failed; after rising about 40 feet the rocket turned and struck the ground 171 feet away. It caused such a fire that neighbors complained to the state fire marshal and Goddard was enjoined from making further tests in Massachusetts.

Ambitious Rocket Tests in Remote Setting

This experience, as well as his personal shyness, led him to seek a more remote setting to conduct his experiments. His ability to shroud his research in mystery was greatly enhanced by Charles A. Lindbergh, fresh from his transatlantic solo flight, who helped Goddard obtain a series of grants from the Guggenheim Fund fostering aeronautical activities. This enabled him to purchase a large tract of desolate land near Roswell, New Mexico, and to set up an independent laboratory to conduct rocket experiments far away from anyone else. Between 1930 and 1941 Goddard carried out more ambitious tests of rocket components in the relative isolation of New Mexico, much of which he summarized in a 1936 study, *Liquid-Propellant Rocket Development*. The culmination of this effort was a successful launch of a rocket to an altitude of 9,000 feet in 1941. In late 1941 Goddard entered naval service and spent the duration of World War II developing a jet-assisted take-off (JATO) rocket to shorten the distance required for heavy aircraft launches. Some of this work led to the development of the throttleable Curtis-Wright XLR25–CW-1 rocket engine that later powered the Bell X-2. Goddard did not live to see this; he died in Annapolis, Maryland, on 10 August 1945.

Goddard's Pioneering Achievements

Goddard accomplished much, but because of his secrecy few people knew about his achievements during his lifetime. These included the following pioneering activities:

- Theorizing on the possibilities of jet-powered aircraft, rocket-borne mail and express, passenger travel in space, nuclear-powered rockets, and journeys to the Moon and other planets (1904–45).
- First mathematical exploration of the practicality of using rockets to reach high altitudes and achieve escape velocity (1912).

- First patent on the idea of multistage rockets (1914).
- First experimental proof that a rocket could provide thrust in a vacuum (1915).
- The basic idea of antitank missiles, developed and demonstrated during work for the U.S. Army in World War I. This was the prototype for the bazooka infantry weapon (1918).
- First publication in the United States of the basic mathematical theory underlying rocket propulsion and space flight (1919).
- First development of a rocket motor burning liquid propellants (1920–26).
- First development of self-cooling rocket motors, variable-thrust rocket motors, practical rocket landing devices, pumps suitable for liquid fuels, and associated components (1920–41).
- First design, construction, and launch of a successful liquid-fueled rocket (1926).
- First development of gyro-stabilization equipment for rockets (1932).
- First use of deflector vanes in the blast of the rocket motor as a method of stabilizing and guiding rockets (1932).

The U.S. government's recognition of Goddard's work came in 1960, when the Department of Defense (DoD) and the NASA awarded his estate $1 million for the use of his patents.

The First Satellite and Animal in Space

By William J. Walter

In this selection William J. Walter tells the story of the launch of the first space satellites. Walter says that in 1950, a small group of American scientists thought that the Earth should be studied from its orbit and that the study should involve an international effort. The scientists invited interested nations to participate in a project known as the International Geophysical Year by launching Earth-orbiting satellites. In 1955 the United States and the Soviet Union accepted the invitation and began building satellites. By 1957 the United States was having trouble with its satellite program but the Soviet Union ushered in the dawn of the space age by secretly launching the *Sputnik I*, the first man-made satellite in human history. News of the Russians' accomplishment stunned and upset Americans. However, four weeks later, the Soviet Union astounded the world again by sending another satellite into orbit, this time carrying the first animal, a dog named Laika. Feeling the pressure to compete, the United States prematurely launched and consequently exploded their first satellite, *Vanguard.* However, in 1958 the United States successfully launched *Explorer I*, a satellite that scientists said had more sophisticated instruments than its Russian counterpart and, indeed, made the first major discovery by a space probe—the existence of radiation belts surrounding the Earth.

William J. Walter is a journalist, documentary filmmaker, and screenwriter. He was former network bureau chief for CNN News and is national programming executive for WQED Television in Pittsburg, Pennsylvania.

I n 1955 scientists throughout the world, particularly in the Soviet Union and the United States, began to focus on a project called the International Geophysical Year (IGY), a watershed event in the Space Age. The origin of the IGY dated back to 1950 when a young astrophysicist named James Van Allen, who had worked with the V-2s [German war rockets] at White Sands, had invited [geophycist] Lloyd Berkner and a few other friends over to his house in Silver Spring, Maryland, for a little soiree. After talking about past international scientific efforts, all of them agreed that it might make sense to coordinate global high-altitude research with scientists from all over the world, and over the coming year their brainstorm eventually evolved into what became the International Geophysical Year. It was to be the first scientific effort aimed at developing a big picture of the way the entire planet worked, and as part of this effort participating nations were invited to launch an Earth-orbiting satellite, an invitation the United States accepted on July 29, 1955. A day later the U.S.S.R. also accepted, and from that moment on the two powers found themselves racing one another into space.

The Soviet Union Has a Satellite Plan

[The Soviet Union's expert jet and rocket engineer Sergei] Korolev was overjoyed with the approval of his satellite plan, and quickly settled on a design that strapped sixteen additional engines to the base of his R-7 [booster rocket formerly used in World War II to launch first intercontinental ballistic missiles (ICBMs)] rocket's fuselage so that it would have enough power to haul what was now being called *sputnik zemli* (travel companion of Earth) into orbit. Korolev didn't delude himself that the satellite was anything very advanced. In fact, it was called PS for Preliminary Satellite and had the look of a strange toy, a silver sphere with four whiplash antennas that swept away from it. Inside was a simple radio transmitter that would begin beeping soon after it was launched. The beep would be flat and standard, like the beep a beacon buoy makes in the water so that you know where it is, which was precisely the idea: The beep would inform Soviet scientists that the little ball had in fact succeeded in circling the planet. Following a few Earthbound preliminary tests, Korolev successfully test-fired two of his newly designed ICBMs out over the Kamchatka peninsula near Japan without incident. By August 1957 his spaceship was ready.

The United States Has Less Success

In the United States, however, success was more elusive. The U.S. satellite project seemed to be sinking beneath the weight of its own intricate design. Even though [German rocket engineer Wernher] von Braun's Redstone rockets [U.S. Army bombardment rockets] already provided a powerful first stage for a satellite launch, the Eisenhower administration had passed them over in favor of the Naval Research Laboratory's more complex Vanguard model rocket, which had been designed purely for civilian and scientific experiments. [President Dwight] Eisenhower felt that it was imperative that the IGY satellite not be tarnished by the military. However, by the summer of 1957 all four of *Vanguard*'s static firing tests had failed, while the Soviets were continually hinting that the launch of *their* satellite was imminent.

By the end of September 1957 Walter Sullivan, science correspondent for *The New York Times*, was pretty certain that what the Soviets were saying was true. Sullivan was covering all aspects of the IGY and had come to Washington to attend a meeting on IGY rocket and satellite plans. The conference was held in the great domed interior of the National Academy of Sciences building beneath grand renaissance-style paintings illustrating man's efforts to comprehend the cosmos. In the spirit of the IGY the Academy had also set up a small demonstration of the Vanguard satellite in the library adjacent to the dome—a full-scale model of the Vanguard rocket and the satellite it would launch. At another display nearby, a tiny animated Vanguard capsule buzzed happily around the planet. This, the little display seemed to say, was the way it would look, after the launch.

Soviets Are Quiet About Launch

Following days of endless descriptions and explanations about Vanguard from American scientists at the conference, everyone was now eagerly awaiting any news the Soviets would have about their launch, but no details were forthcoming except the vague and ubiquitous message that it was . . . imminent. Still hoping to hear some news, however, Sullivan attended a small affair hosted by Russian scientists at the Soviet embassy on the eve of the last day of the conference. By this time Sullivan was so certain a Russian launch would happen soon that he had filed a frontpage story saying as much for the *Times* that would be published on the next day, October 5, 1957.

All of the top American scientists came to the Russian party: Van Allen; William Pickering, head of the Jet Propulsion Laboratory in Pasadena; Herb Friedman, the naval research scientist whose experiments with V-2s had revealed the X-ray universe; Lloyd Berkner; and many others. Each hoped that even here at this social event Anatoli Blagonravov, the head of the Soviet delegation, would come forward with some morsel of information. But Blagonravov remained a block of granite, just as he had all week.

The Soviets Launch Sputnik I

Sullivan had just arrived when a Soviet attaché told him he had a phone call. He excused himself, took the call in the lobby, and was informed by the news desk at the *Times* that they had just received a Reuters wire report saying the Soviets had launched a satellite called *Sputnik I* that was orbiting the planet every ninety-three minutes, beeping away at twenty and forty megacycles.

Sullivan hung up the phone, walked back upstairs, and surveyed the room. As he looked at the faces before him he realized that he was the only person who knew. No one else had a clue, including the Soviets themselves! They had been silent all week long because they didn't have any more information about the launch than anyone else. Either that or they knew when the attempt would be made, but had kept their mouths shut in case the satellite blew up on the launchpad.

The Dawn of the Space Age

After a moment's consideration, Sullivan walked up to a small cluster of Americans that consisted of Richard Porter of General Electric, Pickering, and Berkner, and dropped the news like a hand grenade. Berkner was the first to get his bearings. As the senior American scientist, he clapped his hands, asked for everyone's attention, and raised his glass to toast the Soviets on their accomplishment, the launch of the first man-made satellite in human history—the true dawn of the Space Age. The Soviets quickly shook off their surprise, and then, as Walter Sullivan later recalled, they all smiled "like they had swallowed a thousand canaries." Meanwhile, just down the street at the National Academy, the little animated

Vanguard—the tiny dream ship—whizzed around a fake Earth.

Sergei Korolev waited ninety-three minutes before *Sputnik I* had completed its first orbit and ground control could confirm that it was indeed overhead, beeping. Then he phoned [Russian leader Nikita] Khrushchev at the Kremlin. It was around midnight in Moscow, and Khrushchev had stayed up awaiting the call. He congratulated Korolev and his team, and then went to bed.

The news, however, was not greeted with the same calm equanimity on the other side of the planet.

No one could have anticipated the media riot that followed *Sputnik* in the West—not Eisenhower nor Khrushchev, not the scientists, not anyone. The average American was truly stunned. How could the number-one nation on Earth be caught with its pants wrapped so tightly around its ankles? Yet there *Sputnik* was in orbit, the hard, silvery proof that the Soviet Union actually had bested the United States. Everything from American education to the free-enterprise system was called into question, and newspapers all across America compared the surprise launch to the appalling debacle at Pearl Harbor.

Senate Majority Leader Lyndon Johnson said that the Eisenhower administration had made one of the most monumental political and foreign policy blunders in the history of the nation. *Sputnik*, Johnson reminded America, represented the high ground, mastery of the heavens. Maybe it was all right with others in the government, he told reporters, but he for one didn't care to go to bed by the light of a Communist moon.

The panic was further exacerbated when details about the man who had orchestrated this magnificent celestial maneuver remained cloaked in secrecy. The West was only informed that the man behind *Sputnik* was known as the Chief Designer, and he came to be seen as a Soviet version of the Wizard of Oz, a dark and faceless figure behind the Iron Curtain who rotated the wheels and pulled the levers of the universe at his whim.

A Dog in Space

After seeing the world's reaction to *Sputnik*, Nikita Khrushchev immediately asked his wizard for another demonstration of Communist superiority. Four weeks later, on November 3, 1957, while Washington was still reeling from the first launch, Korolev pulled another lever and launched not only a satellite, but a satellite carrying a dog named Laika inside, and the world was again astounded. The Soviets had launched a living animal into space and were monitoring the effects of the launch as well as radiation and weightlessness on the creature. The mission's success indicated that humans could probably survive orbital flight, which in turn implied that manned spaceflight was precisely the direction in which the Chief Designer was headed.

Vanguard Explodes

Through all of this, *Vanguard* sat at Cape Canaveral, landlocked. Officially it was still in its test stage, but after *Sputnik* the White House announced that the *Vanguard* launch on December 6 would *not* be a test, it would be America's official IGY launch. When the morning of the sixth arrived *Vanguard* sat shimmering in the winter sun by its gantry at Cape Canaveral. It was a beautiful piece of kinetic sculpture, four sleek stages stretching toward the sky. But the truth was that only the first stage had been successfully tested—the rest of it was a crap shoot.

As *Vanguard*'s engines ignited, a mighty flame roared against the launchpad and the rocket rose up four feet, then wavered and paused in time and space under an apocalyptic splash of fire. That was when the full horror of it sunk in: It was not going to fly. Then, very gently, bowing to Isaac Newton's law of gravity, *Vanguard* dropped back onto the launchpad and exploded in a thunderous ball of burning, errant fuel.

When the debacle was over, the launch crew went out to the pad to assess the damage. Blasted, charred metal was scattered everywhere, but to their amazement, among the wreckage they found the third stage laying intact, like a comatose animal. And from inside, the little satellite was innocently whistling away just as though it were in orbit.

More panic ensued. The press called *Vanguard* "Rearguard," "Kaputnik," and "Stayputnik." John McCormack, a member of the House Committee on Science and Astronautics, said that if the United States did not immediately overcome the Soviet Union in the space race, it faced nothing less than national extinction.

However, the United States did have a plan B, and it was already in the works on the day *Vanguard* exploded. At a dinner party on the evening of *Sputnik*'s launch, von Braun had cornered Neil McElroy, the man who was about to become the new Secretary of Defense, and told him that everyone already knew the Russians would succeed first, and that *Vanguard* was never going to make it. Why not give his rocket team a chance? McElroy agreed and granted von Braun permission to prepare a satellite launch as a backup to *Vanguard*. The rocket team hauled out the Redstone, now renamed the Jupiter-C rocket, and prepared a special fourth upper stage to hold the satellite itself. In truth, von Braun had already been working unofficially with Van Allen and Pickering to prepare instruments to fly on his Jupiter-C just in case *Vanguard* failed, and was already well on his way to having a rocket and payload standing by. The project was quickly sanctioned and the satellite was christened *Explorer I*. On January 31, 1958, it was launched into a flawless orbit, and von Braun became a national hero, the man who rose to meet the challenge of the Communist menace in space.

A Major Scientific Discovery

Even though *Explorer I* was not the first or even the second satellite in space, its instruments were far more sophisticated than those on the *Sputniks* and this, at least, led to one first for the United States. The payload had been equipped with an instrument designed by James Van Allen that detected radiation belts encircling the planet which are created when the Earth's magnetic field captures high-energy protons and electrons from the sun, a phenomenon that produces the Auroras at both poles. This was a major scientific discovery, the first, in fact, by a space probe. Most Americans didn't understand the scientific significance of the Van Allen Radiation Belts, but they did perceive in a vague way that the little grapefruits, as Khrushchev liked to call the United States' satellites, could out-perform their Russian counterparts if only U.S. rockets could manage to get them into space.

The Formation of the National Aeronautics and Space Administration

By Stephen J. Garber and Roger D. Launius

Following World War II, the United States and the Soviet Union were engaged in the Cold War where tensions spilled over into the field of space exploration. In what was envisioned as a cooperative international scientific effort to explore Earth with space satellites, the United States and Russia became staunch competitors in a space race. The U.S. government and public considered it a national crisis when the Soviet Union was the first to launch a space satellite, *Sputnik I*, into Earth's orbit. As a result, in October 1958, Congress and President Dwight Eisenhower created the National Aeronautics and Space Administration (NASA), bringing more money and expertise into the field of U.S. space exploration. NASA conducted space missions within the first months of its creation, and in its formative years the program was busy with several major projects, including Project Mercury. As part of Project Mercury, Alan B. Shepard Jr. became the first American in space in 1961, followed by John H. Glenn in 1962, the first American astronaut to orbit Earth. Project Gemini followed Project Mercury and consisted of ten space flights. Edward H. White Jr.'s spacewalk in 1965 highlighted the program and was another first in U.S. spaceflight history.

Stephen J. Garber is a policy analyst in the NASA history office. He has written several articles on aerospace history topics including

Stephen J. Garber and Roger D. Launius, "A Brief History of the National Aeronautics and Space Administration," www.hq.nasa.gov.

one on the design of the space shuttle and the congressional cancellation of NASA's Search for Extraterrestrial Intelligence program. Roger D. Launius is chief historian for NASA and curator in the Division of Space History at the Smithsonian Institute in Washington, D.C. He has lectured widely and is vice president of publications and editor of *Space Times: The Magazine of the American Astronautical Society.*

"**A**n Act to provide for research into the problems of flight within and outside the Earth's atmosphere, and for other purposes." With this simple preamble, the Congress and the President of the United States created the national Aeronautics and Space Administration (NASA) on October 1, 1958. NASA's birth was directly related to the pressures of national defense. After World War II, the United States and the Soviet Union were engaged in the Cold War, a broad contest over the ideologies and allegiances of the nonaligned nations. During this period, space exploration emerged as a major area of contest and became known as the space race.

During the late 1940s, the Department of Defense pursued research and rocketry and upper atmospheric sciences as a means of assuring American leadership in technology. A major step forward came when President Dwight D. Eisenhower approved a plan to orbit a scientific satellite as part of the International Geophysical Year (IGY) for the period, July 1, 1957 to December 31, 1958, a cooperative effort to gather scientific data about the Earth. The Soviet Union quickly followed suit, announcing plans to orbit its own satellite.

The Naval Research Laboratory's Project Vanguard was chosen on 9 September 1955 to support the IGY effort, largely because it did not interfere with high-priority ballistic missile development programs. It used the non-military Viking rocket as its basis while an Army proposal to use the Redstone ballistic missile [U.S. Army bombardment rocket] as the launch vehicle waited in the wings. Project Vanguard enjoyed exceptional publicity throughout the second half of 1955, and all of 1956, but the technological demands upon the program were too great and the funding levels too small to ensure success.

A full-scale crisis resulted on October 4, 1957 when the Soviets launched *Sputnik I*, the world's first artificial satellite as its IGY entry. This had a "Pearl Harbor" effect on American public opinion, creating an illusion of a technological gap and provided the impetus for increased spending for aerospace endeavors, technical and scientific educational programs, and the chartering of new federal agencies to manage air and space research and development.

Explorer 1 and the Discovery of Radiation Zones

More immediately, the United States launched its first Earth satellite on January 31, 1958, when *Explorer 1* documented the existence of radiation zones encircling the Earth. Shaped by the Earth's magnetic field, what came to be called the Van Allen Radiation Belt, these zones partially dictate the electrical charges in the atmosphere and the solar radiation that reaches Earth. The U.S. also began a series of scientific missions to the Moon and planets in the latter 1950s and early 1960s.

A direct result of the Sputnik crisis, NASA began operations on October 1, 1958, absorbing into itself the earlier National Advisory Committee for Aeronautics intact: its 8,000 employees, an annual budget of $100 million, three major research laboratories—Langley Aeronautical Laboratory, Ames Aeronautical Laboratory, and Lewis Flight Propulsion Laboratory—and two smaller test facilities. It quickly incorporated other organizations into the new agency, notably the space science group of the Naval Research Laboratory in Maryland, the Jet Propulsion Laboratory managed by the California Institute of Technology for the Army, and the Army Ballistic Missile Agency in Huntsville, Alabama, where [German rocket engineer] Wernher von Braun's team of engineers were engaged in the development of large rockets. Eventually NASA created other Centers and today it has ten located around the country.

NASA's First Major Programs

NASA began to conduct space missions within months of its creation, and during its first twenty years NASA conducted several major programs:

- Human space flight initiatives—Mercury's single astronaut program (flights during 1961–1963) to ascertain if a human could survive in space; Project Gemini (flights during 1965–

1966) with two astronauts to practice space operations, especially rendezvous and docking of spacecraft and extravehicular activity (EVA); and Project Apollo (flights during 1968–1972) to explore the Moon.

- Robotic missions to the Moon (Ranger, Surveyor, and Lunar Orbiter), Venus (*Pioneer Venus*), Mars (*Mariner 4, Viking 1* and *2*), and the outer planets (*Pioneer 10* and *11, Voyager 1* and *2*).
- Aeronautics research to enhance air transport safety, reliability, efficiency, and speed (X-15 hypersonic flight, lifting body flight research, avionics and electronics studies, propulsion technologies, structures research, aerodynamics investigations).
- Remote-sensing Earth satellites for information gathering (Landsat satellites for environmental monitoring).
- Applications satellites for communications (*Echo 1, TIROS*, and *Telstar*) and weather monitoring.
- An orbital workshop for astronauts, *Skylab.*
- A reusable spacecraft for traveling to and from Earth orbit, the Space Shuttle.

Early Spaceflights: Mercury and Gemini

NASA's first high-profile program involving human spaceflight was Project Mercury, an effort to learn if humans could survive the rigors of spaceflight. On May 5, 1961, Alan B. Shepard Jr. became the first American to fly into space, when he rode his Mercury capsule on a 15-minute suborbital mission. John H. Glenn Jr. became the first U.S. astronaut to orbit the Earth on February 20, 1962. With six flights, Project Mercury achieved its goal of putting piloted spacecraft into Earth orbit and retrieving the astronauts safely.

Project Gemini built on Mercury's achievements and extended NASA's human spaceflight program to spacecraft built for two astronauts. Gemini's 10 flights also provided NASA scientists and engineers with more data on weightlessness, perfected reentry and splashdown procedures, and demonstrated rendezvous and docking in space. One of the highlights of the program occurred during Gemini 4, on June 3, 1965, when Edward H. White Jr., became the first U.S. astronaut to conduct a spacewalk.

CHAPTER 2

Problems and Successes of Putting Humans in Space

The First Human in Space

By Peter Bond

In April 1961, Russian cosmonaut Yuri Gagarin became the first human to fly in space. In this selection, Peter Bond relates the story of Gagarin's early life and training as a jet pilot in the Soviet Union. Bond says that the twenty-seven-year-old air force lieutenant was unanimously selected by the Soviet state commission to man the momentous flight. Bond details Gagarin's experiences leading up to and during the liftoff, as well as his experiments with eating, drinking, and being weightless for the first time while traveling in space. The *Vostok* spacecraft that Gagarin traveled in was automatically piloted from launch to touchdown and reached an unprecedented speed of eighteen thousand miles per hour. After one trip around Earth in a mere one hour and eighteen minutes, Gagarin disappointedly headed back for a landing as planned. Gagarin had become a legend and was exceedingly happy that his beloved homeland was the first to reach outer space and that he had been the one chosen to fly the mission. However, this was the only spaceflight the Russian cosmonaut made due to his premature death in a 1968 plane crash.

Peter Bond is a press officer for the Royal Astronomical Society, and a consultant for the European Space Agency and the Particle Physics and Astronomy Research Council. Bond is a fellow of both the British Interplanetary Society and the Royal Astronomical Society. He has written hundreds of articles on space and astronomy for British and American newspapers and magazines. He is also the author of six books.

At 10.55 AM Moscow time on 12 April 1961, a man wearing an orange flight suit and a white pressure helmet landed in a field watched only by a cow and two bemused peasants.

Peter Bond, *Heroes in Space: From Gagarin to* Challenger. Oxford, UK: Basil Blackwell, 1987.

As he pulled himself together and staggered to his feet, he saw the woman and girl staring at him as he unhitched his parachute. Mrs Anya Takhtarova stepped towards him and doubtfully stammered, 'Have you come from outer space?' to which the young man triumphantly replied, 'Yes. Would you believe it? I certainly have.' The poor woman must have looked so frightened that he hastily added, 'Don't be alarmed . . . I'm Soviet!' He then walked over with them to inspect the charred space capsule which was still smouldering nearby. They were soon joined by excited tractor drivers who had been working near the landing site. Some of them had been listening to the radio and so knew the identity of the stranger. As they helped him remove the protective suit, they jubilantly shouted his name, 'Yuri Gagarin! Yuri Gagarin!' Someone gave him a cap to wear during the short interval before a helicopter brought members of the spacecraft landing support team. After a brief check-up, the exhilarated but weary cosmonaut was whisked away to the nearby city of Saratov for a period of rest and medical observation. Two days later, Gagarin was the centrepiece of a magnificent hero's welcome in Red Square, Moscow, which was televised live all over Europe, and later broadcast to the rest of the world.

Today, the landing site at Smelovaka, near Saratov, is marked by a 130 ft (40 m) high titanium obelisk, and thousands of visitors travel to the flat, fertile steppes to pay homage to the first human who travelled through space. His unique achievement makes Gagarin's place in history assured, despite the fact that he never flew in space again. No matter what miracles of human endurance or technological excellence the future may hold, Yuri Gagarin was the first; the pioneer who entered the unknown and opened the gate for his successors.

Yuri Gagarin's Background

The man who made his mark on world history was born in a wooden house in the small village of Klushino, near Smolensk in the western Soviet Union. His father was a carpenter, and times were hard during the Nazi occupation. During the war, the young Yuri saw an aeroplane for the first time. The family moved in 1945 to the town of Gzhatsk (now named Gagarin) where Yuri completed school and went to a trade-training school, then worked in a foundry. He was sent to an industrial training college in Saratov and there he joined an aero club where he gained experience in solo flight and parachute jumping. Gagarin then progressed to flying jets at the air force training school in Orenburg, from which he graduated in 1957. The final step came in the spring of 1960 when his application was accepted to join a small group of airmen who were to be trained as future cosmonauts. By this time, he had clocked a mere 230 hours in the air.

There followed a year of intensive training at the specially constructed cosmonaut training centre, Zvezdniy Gorodok (Star City). Prophetically, perhaps, Gagarin was the first of the group of 20 to sit inside the Vostok ship on the day chief designer Korolev introduced it to them. However, it was not until 8 April 1961 that the 27-year-old air force lieutenant Gagarin was selected by the state commission for the momentous flight which was to take place only four days later. Apparently, Gagarin was nominated by General Nikolai Kamanin, head of the cosmonaut team, and seconded by Korolev with the words, 'I find in him an analytical mind and rare industriousness. We need profound information about outer space and I have no doubt that Gagarin will bring it.' The vote for Gagarin was unanimous.

Preparing for the Flight

The morning of 11 April saw the rocket assembled and transported in a horizontal position along the rail tracks from the flat-roofed assembly building to the launch pad. There it was raised to a vertical position, and held in the grip of four supporting arms. Maintenance technicians and service engineers busied themselves as the main rocket and the four strap-on boosters were loaded with liquid oxygen and kerosene, and all the mechanical, electrical, communications and life-support systems were checked and rechecked. Amidst this feverish activity, Yuri Gagarin was introduced to the service personnel and greeted by rapturous applause. Afterwards, Gagarin and Korolev spent over an hour alone at the top of the rocket beside the tiny Vostok capsule.

There were three identical wooden houses, surrounded by young poplar trees, only 15 minutes drive from the launch pad. It was there that the main participants in the historic events to follow spent a fairly restless night. Korolev's heart condition was causing him pain and as he opened a window he noticed that Konstantin Rudnev, head of the state commission, could not sleep either—the lights in both the neighbouring houses were shining brightly. The two men met in the garden and walked around discussing the launch scheduled for six hours time. Korolev even crept to the door of Gagarin's room and peered in, but both he and his back-up, Titov, seemed sound asleep, despite the biosensors attached by doctors to monitor their responses to stress. They apparently had little difficulty, however, in being roused by Yevgeni Karpov, Kamanin's lieutenant, at 5.30 on the morning of 12 April.

After breakfast, Gagarin and Titov were helped into the 'space wardrobe': sensors were attached to their bodies to monitor their physical condition while in flight, the woollen undergarment was put on, then the pressure suit to protect against radiation and failure of the heating system, then the orange coveralls to aid spotting on landing, and finally the gloves, boots and helmet. For the time being, the transparent visor was kept open as the two men walked awkwardly from the cottage door to the waiting bus. Within a few minutes they were at the Baikonur space centre, and beneath the towering rocket. Gagarin declared himself fit and ready to the state commission.

A Moving Farewell

The final act was a moving farewell speech from the platform at the foot of the lift, addressed to 'people of all countries and continents'. Gagarin reflected on his role in the forthcoming mission:

> At this instant, the whole of my life seems to be condensed into one wonderful moment. Everything that I have experienced, everything that I have done hitherto, was experienced and done for the sake of this moment. . . . Of course I am happy. In all times and epochs the greatest happiness for man has been to take part in new discoveries. To be the first to enter the cosmos, to engage, singlehanded, in an unprecedented duel with nature—could one dream of anything more!

Gagarin then dedicated the flight to 'the people of a communist society' and bade the emotional audience farewell.

Yuri Gagarin raised his hands aloft then turned and entered the lift. On leaving the lift at the top, he climbed a short metal ladder to the platform which led to the Vostok cabin. He was helped through the hatch and into the specially designed couch, then the hatch was closed and Gagarin was left to contemplate his fate. As the hour dragged by, some music was played over the intercom to help pass the time. After a short delay due to a faulty valve, the final countdown began. The last commands were issued: 'Switch to "go" position!' Gagarin settled back in the reclining seat. 'Air purging! Idle run!' The fuel tower slowly withdrew from alongside the rocket. 'Ignition!' The arm carrying the power cable swung away. 'Lift-off!' At 9.07 AM Moscow time the supporting arms gently opened like the petals of a flower, and the huge white rocket rose, imperceptibly at first, into the clear blue sky. 'Off we go!' came the jubilant voice of Gagarin as his ship rapidly disappeared from view, leaving only a trail of smoke and flame.

The First Views of Earth

The world's first cosmonaut was pressed harder and harder into his padded seat, unable to move, just as he had experienced in the centrifuge during his training. As the rocket broke the sound barrier, Gagarin lay back in an eerie quiet, broken only by the separation of the four strap-on boosters. Korolev listened worriedly to the buzzing on the intercom, then sighed with relief as Gagarin reported, 'The fairing has been discarded. . . . I see the Earth. The loads are increasing. Feeling fine.' Pressure reached about 6G as the second stage boosted him into orbit before shutting off and throwing him suddenly forwards. This time, however, he did not rebound into the seat as the straps held him back, but, instead, felt himself suspended above the couch as everything not fastened down began to float past him. The second stage separated and Gagarin got down to work, noting instrument readings, checking equipment and recording the effects of weightlessness and the appearance of his surroundings. 'The sky looks very, very dark and the Earth is bluish.' Later he reported,

> The sunlit side of the Earth is visible quite well, and one can easily distinguish the shores of continents, islands and great rivers, large areas of water and folds of the land. Over Russia I saw distinctly the big squares of collective-farm fields, and it was possible to distinguish which was ploughed land and which was meadow. During the flight I saw for the first time, with my own eyes, the Earth's spherical shape.

Soon after Vostok entered Earth orbit, Moscow Radio proudly

announced to the world: 'On April 12 1961, in the Soviet Union, the Vostok, the world's first manned spacecraft, has been launched into orbit around the Earth. . . .' Crowds began to gather in Red Square, cheering, embracing, holding placards. Meanwhile, Gagarin was watching the Earth flash by as he travelled at the unprecedented speed of nearly 18,000 mph (29,000 kph) over Siberia, Japan, Cape Horn and back towards Africa. The only moment of panic on the ground came while Gagarin was out of direct radio contact; the planners opted for brevity and clarity so a number code was used with 'five' indicating that all was well. Suddenly the machine began repeating 'threes' and everyone in the control room turned pale as they stared at the tape. Then the numbers changed again . . . 'fives' once more! Korolev dropped heavily into a chair and sighed, 'It's seconds like that which shorten a designer's life.'

Gagarin flew on, unaware of the emotional scenes on the ground. He happily practised eating and drinking from a supply provided in a small container by his right shoulder. Weightlessness he found very relaxing, giving a sense of increased room in the cramped cabin. He reported, 'Handwriting did not change, though the hand was weightless. But it was necessary to hold the writing block, as otherwise it would float from the hands.' He was not needed to pilot the craft, however, since it was designed to operate automatically from launch to touch-down. It seems that the original intention of the designers was to use the cosmonauts as passive passengers and so no manual control was provided. The cosmonauts, all experienced pilots, objected to this, and so a manual back-up was added. To ensure the system was not inadvertently activated, a combination lock was fitted with the numbers (1-4-5 in Gagarin's case) kept in an envelope attached to the cabin wall.

Heading Back to Earth

The automatic system operated on schedule, and Gagarin later admitted a sense of disappointment as the craft swivelled round ready for retro-fire. As Vostok flew over East Africa, only 1 hour 18 minutes after launch, the retro-rockets fired to brake the capsule. Had they failed, the orbit was such that friction with the upper atmosphere would have returned the craft to Earth within ten days. The rocket section was safely jettisoned, and Gagarin began re-entry, facing backwards as so many other cosmonauts and astronauts were to do in the years to come. Through the portholes he was able to witness the terrifying firework display as the craft's exterior heated to thousands of degrees and the protective coating burned away. The acceleration forces built up to more than 8G before a small drogue parachute was deployed through a hatch at a height of 2½ ml (4 km) above the ground. This was followed at 1½ ml (2½ km) by the main parachute which slowed the steel capsule before its bumpy landing.

Soviets Maintain Secrecy

The official Soviet version of events was that Gagarin landed inside the charred capsule, but this has been questioned by a number of Western experts who believe that he ejected from the Vostok at about 20,000 feet (6 km) and landed using his own personal parachute. Certainly this was the method used by all subsequent Vostok crews. Was this because Gagarin's landing in the capsule was so rough, or was it a method common to all Vostoks, including Gagarin's? The Soviet motive for hiding the truth seems obscure, although it has been suggested that the flight would not have been registered as a record by the Federation Aeronautique Internationale, the organization which certifies all flights records, if Gagarin was known to have landed separately from his craft.

Soviet secrecy covered most aspects of the flight and only today is it possible to piece together the whole story. It was four years before any photographs of the Vostok were released, by which time the craft was obsolete, while the launch vehicle was not seen in the West until the Paris Air Show of May 1967. Inevitably this secretiveness by the Soviet authorities led to wild speculation and rumours in Western newspapers concerning the Soviet space programme. The *New York Herald Tribune* of 19 April 1961 even published a headline, 'Did Gagarin Do It?', alleging that a dummy may have been in the capsule. British newspapers quoted a Mr Bobrovsky who claimed 'reliable sources' for a story that a test pilot, Sergei Ilyushin, son of the famous Soviet aircraft designer, was launched into space a few days before Gagarin but returned to earth after three orbits suffering from loss of balance and was rushed unconscious to hospital. Other stories circulated concerning alleged cosmonaut deaths: some were said to have died in suborbital flights, others in rocket explosions and launch failures, yet others were supposedly trapped in Earth orbit.

In 1986, the Soviet media specifically denied all of these allegations, although they did admit the death of one cosmonaut during a training accident on 23 March 1961. Valentin Bondarenko, the youngest member of the group, died as the result of a fire in an isolation chamber. The flames spread rapidly in the oxygen-rich atmosphere. Protected only by his woollen training suit, Bondarenko was so badly burned that the doctors were unable to save him. He was buried in Kharkov, his birthplace.

A Legend in His Lifetime

Yuri Gagarin became a legend in his own lifetime. The small, friendly, unassuming man with the boyish smile travelled the world as a roving ambassador, met the world's leaders and received the highest honours and decorations. He became a deputy of the Supreme Soviet, representing the Smolensk region, and always continued to help and advise his fellow cosmonauts. He graduated with honours from the Zhukovsky Air Force Academy and became Commander of the Cosmonauts' Detachment. Eventually he seems to have tired of the life of a celebrity. As he put it, 'Being a cosmonaut is my profession, and I did not choose it just to make the first flight and then give it up.' He went back into training and was appointed back-up pilot to Vladimir Komarov for the first flight of the new Soyuz craft in 1967. A year later, Gagarin too was dead. He was killed, along with his training instructor, in a plane crash on 27 March 1968. His ashes were buried in the Kremlin Wall, alongside other Soviet heroes. Some seven years earlier, he had stood in Red Square and spoken to the assembled multitudes: 'One can say with assurance that on Soviet spacecraft we will fly even over more distant routes. I am boundlessly happy that my beloved homeland was the first to accomplish this flight, was the first to reach outer space.'

Vowing to Put a Man on the Moon

By John F. Kennedy

This famous speech was delivered by President John F. Kennedy in a special address to Congress on May 25, 1961. It was a time of dramatic events in space exploration; the Soviets had recently put the first human into space, and within a month, the United States had done the same. President Kennedy was anxious for the United States to take a clear lead in space exploration. In his address, the president proposed new national goals to advance the U.S. space program. The most outstanding of these goals was to have a man land on the moon and safely return before the end of the decade. The president stressed the national importance of achieving this unprecedented and impressive feat. He acknowledged the seriousness of the commitment to this goal as well, which would take many years and demand major funding to accomplish. Nevertheless, the president implored Congress and the people of the United States to make a firm commitment to support his pivotal plans that would make the United States the leader in space exploration. President John F. Kennedy was the thirty-fifth president of the United States and was in office from January 1961 to November 1963.

I f we are to win the battle that is now going on around the world between freedom and tyranny, the dramatic achievements in space which occurred in recent weeks[1] should have made clear to us all, as did the Sputnik in 1957, the impact of this adventure on the minds of men everywhere, who are attempting to make a determination of which road they should take. Since early in my term, our efforts in space have been under review. With the advice

---

1. In April/May 1961 the Soviets put the first human into space. The Americans followed suit a few weeks later.

John F. Kennedy, address to the U.S. Congress, Washington, DC, May 25, 1961.

of the Vice President, who is Chairman of the National Space Council, we have examined where we are strong and where we are not. Now it is time to take longer strides—time for a great new American enterprise—time for this nation to take a clearly leading role in space achievement, which in many ways may hold the key to our future on Earth.

I believe we possess all the resources and talents necessary. But the facts of the matter are that we have never made the national decisions or marshaled the national resources required for such leadership. We have never specified long-range goals on an urgent time schedule, or managed our resources and our time so as to insure their fulfillment.

Recognizing the head start obtained by the Soviets with their large rocket engines, which gives them many months of lead-time, and recognizing the likelihood that they will exploit this lead for some time to come in still more impressive successes, we nevertheless are required to make new efforts on our own. For while we cannot guarantee that we shall one day be first, we can guarantee that any failure to make this effort will be our last. We take an additional risk by making it in full view of the world, but as shown by the feat of astronaut [Alan] Shepard [the first American astronaut to go into space] this very risk enhances our stature when we are successful. But this is not merely a race. Space is open to us now; and our eagerness to share its meaning is not governed by the efforts of others. We go into space because whatever mankind must undertake, free men must fully share.

Land a Man on the Moon

I therefore ask the Congress, above and beyond the increases I have earlier requested for space activities, to provide the funds which are needed to meet the following national goals:

First, I believe that this nation should commit itself to achieving the goal, before this decade is out, of landing a man on the Moon and returning him safely to the Earth. No single space project in this period will be more impressive to mankind, or more important for the long-range exploration of space; and none will be so difficult or expensive to accomplish. We propose to accelerate the development of the appropriate lunar space craft. We propose to develop alternate liquid and solid fuel boosters, much larger than any now being developed, until certain which is superior. We pro-

President John F. Kennedy addresses Congress in May 1961, vowing that the United States will put a man on the moon by the end of the decade.

pose additional funds for other engine development and for unmanned explorations—explorations which are particularly important for one purpose which this nation will never overlook: the survival of the man who first makes this daring flight. But in a very real sense, it will not be one man going to the Moon—if we make this judgment affirmatively, it will be an entire nation. For all of us must work to put him there.

Secondly, an additional 23 million dollars, together with 7 million dollars already available, will accelerate development of the Rover nuclear rocket. This gives promise of some day providing a means for even more exciting and ambitious exploration of space, perhaps beyond the Moon, perhaps to the very end of the solar system itself.

Third, an additional 50 million dollars will make the most of our present leadership, by accelerating the use of space satellites for world-wide communications.

Fourth, an additional 75 million dollars—of which 53 million dollars is for the Weather Bureau—will help give us at the earliest possible time a satellite system for world-wide weather observation.

Let it be clear—and this is a judgment which the Members of

the Congress must finally make—let it be clear that I am asking the Congress and the country to accept a firm commitment to a new course of action—a course which will last for many years and carry very heavy costs: 531 million dollars in fiscal '62—an estimated seven to nine billion dollars additional over the next five years. If we are to go only half way, or reduce our sights in the face of difficulty, in my judgment it would be better not to go at all.

Now this is a choice which this country must make, and I am confident that under the leadership of the Space Committees of the Congress, and the Appropriating Committees, that you will consider the matter carefully.

It is a most important decision that we must make as a nation. But all of you have lived through the last four years and have seen the significance of space and the adventures in space, and no one can predict with certainty what the ultimate meaning will be of mastery of space.

I believe we should go to the Moon. But I think every citizen of this country as well as the Members of the Congress should consider the matter carefully in making their judgment, to which we have given attention over many weeks and months, because it is a heavy burden, and there is no sense in agreeing or desiring that the United States take an affirmative position in outer space, unless we are prepared to do the work and bear the burdens to make it successful. If we are not, we should decide today and this year.

A Major National Commitment

This decision demands a major national commitment of scientific and technical manpower, material and facilities, and the possibility of their diversion from other important activities where they are already thinly spread. It means a degree of dedication, organization and discipline which have not always characterized our research and development efforts. It means we cannot afford undue work stoppages, inflated costs of material or talent, wasteful interagency rivalries, or a high turnover of key personnel.

New objectives and new money cannot solve these problems. They could in fact, aggravate them further—unless every scientist, every engineer, every serviceman, every technician, contractor, and civil servant gives his personal pledge that this nation will move forward, with the full speed of freedom, in the exciting adventure of space.

The Significance of the Apollo Program

By Carl Sagan

In the following narrative, astronomer Carl Sagan explains the many significant contributions of the Apollo space program. Sagan says the Apollo missions inspired optimism about technology and conveyed a confidence, energy, and breadth of vision that captured the imagination of the world. However, Sagan believes that President John F. Kennedy launched the Apollo program with a political agenda in mind rather that a scientific one. In the early 1960s when Cold War tensions had grown between the United States and the Soviet Union, U.S. military leaders believed that the nation that controlled space would also control Earth. Therefore, the Apollo mission's goal to put a man on the moon was, in part, an attempt to gain leadership credibility and prestige. Sagan says that the Apollo program easily acquired the necessary funding and ironically supplanted all the numerous U.S. military space plans at the time, shifting the U.S.-Soviet space race from a military to a civilian pursuit. Among the remarkable technological and scientific advances made in the seventeen Apollo missions, *Apollo 11*'s landing of the first man on the moon in 1969 was a turning point in history. Humankind had escaped the bounds of Earth and touched into a mythological realm. Sagan says the moon mission, as well as the other Apollo missions, expanded scientists' knowledge of the composition, age, and history of the moon, but more importantly, if not for the Apollo missions, other exploration and discoveries throughout the solar system may not have occurred.

Carl Sagan, who died in 1996, was a professor of astronomy and space sciences and director of the Laboratory for Planetary Studies at Cornell University in New York City. Sagan played a leading role

Carl Sagan, *Pale Blue Dot*. New York: Random House, 1994. Copyright © 1994 by Carl Sagan. All rights reserved. Reproduced by permission of the publisher.

in the *Mariner, Viking*, and *Voyager* spacecraft expeditions. He was the cofounder and first president of the Planetary Society, the largest space interest organization in the world, created to further space exploration and the search for extraterrestrial life. He received numerous awards including the Pulitzer Prize, and the NASA Medals for Exceptional Scientific Achievement and for Distinguished Public Service. Sagan authored several books including *Cosmos*, the best-selling science book ever published in the English language.

Of all the events surrounding *Apollo 11*'s landing on the Moon on July 20, 1969, my most vivid recollection is its unreal quality. Neil Armstrong and Buzz Aldrin shuffled along the gray, dusty lunar surface, the Earth looming large in their sky, while Michael Collins, now the Moon's own moon, orbited above them in lonely vigil. Yes, it was an astonishing technological achievement and a triumph for the United States. Yes, the astronauts displayed death-defying courage. Yes, as Armstrong said as he first alighted, this was a historic step for the human species. But if you turned off the byplay between Mission Control and the Sea of Tranquility, with its deliberately mundane and routine chatter, and stared into that black-and-white television monitor, you could glimpse that we humans had entered the realm of myth and legend. . . .

Reaching the Unattainable

The Moon was a metaphor for the unattainable: "You might as well ask for the Moon," they used to say. Or "You can no more do that than fly to the Moon." For most of our history, we had no idea what it was. A spirit? A god? A thing? It didn't look like something big far away, but more like something small nearby—something the size of a plate, maybe, hanging in the sky a little above our heads. Ancient Greek philosophers debated the proposition "that the Moon is exactly as large as it looks" (betraying a hopeless confusion between linear and angular size). *Walking* on the Moon would have seemed a screwball idea; it made more sense to imagine somehow climbing up into the sky on a ladder or on the back of a giant bird, grabbing the Moon, and bringing it down to Earth. Nobody ever succeeded, although there were myths aplenty about heroes who had tried.

Not until a few centuries ago did the idea of the Moon as a *place*, a quarter-million miles away, gain wide currency. And in that brief flicker of time, we've gone from the earliest steps in understanding the Moon's nature to walking and joy-riding on its surface. We calculated how objects move in space; liquefied oxygen from the air; invented big rockets, telemetry, reliable electronics, inertial guidance, and much else. Then we sailed out into the sky.

I was lucky enough to be involved in the *Apollo* program, but I don't blame people who think the whole thing was faked in a Hollywood movie studio. In the late Roman Empire, pagan philosophers had attacked Christian doctrine on the ascension to Heaven of the body of Christ and on the promised bodily resurrection of the dead—because the force of gravity pulls down all "earthly bodies." St. Augustine rejoined: "If human skill can by some contrivance fabricate vessels that float, out of metals which sink . . . how much more credible is it that God, by some hidden mode of operation, should even more certainly effect that these earthly masses be emancipated" from the chains that bind them to Earth? That *humans* should one day discover such a "mode of operation" was beyond imagining. Fifteen hundred years later, we emancipated ourselves.

The achievement elicited an amalgam of awe and concern. Some remembered the story of the Tower of Babel. Some, orthodox Moslems among them, felt setting foot on the Moon's surface to be impudence and sacrilege. Many greeted it as a turning point in history.

The Moon is no longer unattainable. A dozen humans, all Americans, have made those odd bounding motions they called "moonwalks" on the crunchy, cratered, ancient gray lava—beginning on that July day in 1969. But since 1972, no one from any nation has ventured back. Indeed, none of us has gone *anywhere* since the glory days of *Apollo* except into low Earth orbit—like a toddler who takes a few tentative steps outward and then, breathless, retreats to the safety of his mother's skirts.

Once upon a time, we soared into the Solar System. For a few years. Then we hurried back. Why? What happened? What was *Apollo* really about?

An Audacious Political Plan

The scope and audacity of John Kennedy's May 25, 1961, message to a joint session of Congress on "Urgent National Needs"— the speech that launched the *Apollo* program—dazzled me. We would use rockets not yet designed and alloys not yet conceived, navigation and docking schemes not yet devised, in order to send a man to an unknown world—a world not yet explored, not even in a preliminary way, not even by robots—and we would bring him safely back, and we would do it before the decade was over. This confident pronouncement was made before any American had even achieved Earth orbit.

As a newly minted Ph.D., I actually thought all this had something centrally to do with science. But the President did not talk about discovering the origin of the Moon, or even about bringing samples of it back for study. All he seemed to be interested in was sending someone there and bringing him home. It was a kind of *gesture.* Kennedy's science advisor, Jerome Wiesner, later told me he had made a deal with the President: If Kennedy would not claim that *Apollo* was about science, then he, Wiesner, would support it. So if not science, what?

The *Apollo* program is really about politics, others told me. This sounded more promising. Nonaligned nations would be tempted to drift toward the Soviet Union if it was ahead in space exploration, if the United States showed insufficient "national vigor." I didn't follow. Here was the United States, ahead of the Soviet Union in virtually every area of technology—the world's economic, military, and, on occasion, even moral leader—and Indonesia would go Communist because Yuri Gagarin beat John Glenn to Earth orbit? What's so special about space technology? Suddenly I understood.

The New High Ground

Sending people to orbit the Earth or robots to orbit the Sun requires rockets—big, reliable, powerful rockets. Those same rockets can be used for nuclear war. The same technology that transports a man to the Moon can carry nuclear warheads halfway around the world. The same technology that puts an astronomer and a telescope in Earth orbit can also put up a laser "battle station." Even back then, there was fanciful talk in military circles, East and West, about space as the new "high ground," about the nation that "controlled" space "controlling" the Earth. Of course strategic rockets were already being tested on Earth. But heaving a ballistic missile with a dummy warhead into a target zone in the middle of the Pacific Ocean doesn't buy much glory. Sending people into space captures the attention and imagination of the world.

You wouldn't spend the money to launch astronauts for this reason alone, but of all the ways of demonstrating rocket potency, this one works best. It was a rite of national manhood; the shape of the boosters made this point readily understood without anyone actually having to explain it. The communication seemed to be transmitted from unconscious mind to unconscious mind without the higher mental faculties catching a whiff of what was going on. . . .

Apollo Supplants Military Programs

When President Kennedy formulated the *Apollo* program, the Defense Department had a slew of space projects under development—ways of carrying military personnel up into space, means of conveying them around the Earth, robot weapons on orbiting

In 1971 James Irwin and fellow astronauts aboard Apollo 15 *explored the Moon's surface with the first lunar rover.*

platforms intended to shoot down satellites and ballistic missiles of other nations. *Apollo* supplanted these programs. They never reached operational status. A case can be made then that *Apollo* served another purpose—to move the U.S.-Soviet space competition from a military to a civilian arena. There are some who believe that Kennedy intended *Apollo* as a substitute for an arms race in space. Maybe.

For me, the most ironic token of that moment in history is the plaque signed by President Richard M. Nixon that *Apollo 11* took to the Moon. It reads: "We came in peace for all mankind." As the United States was dropping 7½ megatons of conventional explosives on small nations in Southeast Asia, we congratulated ourselves on our humanity: We would harm no one on a lifeless rock. That plaque is there still, attached to the base of the *Apollo 11* Lunar Module, on the airless desolation of the Sea of Tranquility. If no one disturbs it, it will still be readable a million years from now.

Six more missions followed *Apollo 11*, all but one of which successfully landed on the lunar surface. *Apollo 17* was the first to carry a scientist. As soon as he got there, the program was canceled. The first scientist and the last human to land on the Moon were the same person. The program had already served its purpose that July night in 1969. The half-dozen subsequent missions were just momentum.

Apollo was not mainly about science. It was not even mainly about space. *Apollo* was about ideological confrontation and nuclear war—often described by such euphemisms as world "leadership" and national "prestige." Nevertheless, good space science was done. We now know much more about the composition, age, and history of the Moon and the origin of the lunar landforms. We have made progress in understanding where the Moon came from. Some of us have used lunar cratering statistics to better understand the Earth at the time of the origin of life. But more important than any of this, *Apollo* provided an aegis, an umbrella under which brilliantly engineered robot spacecraft were dispatched throughout the Solar System, making that preliminary reconnaissance of dozens of worlds. The offspring of *Apollo* have now reached the planetary frontiers.

If not for *Apollo*—and, therefore, if not for the political purpose it served—I doubt whether the historic American expeditions of exploration and discovery throughout the Solar System would have occurred. The *Mariners, Vikings, Pioneers, Voyagers*, and

Galileo are among the gifts of *Apollo, Magellan* and *Cassini* are more distant descendants. Something similar is true for the pioneering Soviet efforts in Solar System exploration, including the first soft landings of robot spacecraft—*Luna 9, Mars 3, Venera 8*—on other worlds.

An Enthusiasm for the Future

Apollo conveyed a confidence, energy, and breadth of vision that did capture the imagination of the world. That too was part of its purpose. It inspired an optimism about technology, an enthusiasm for the future. If we could fly to the Moon, as so many have asked, what else were we capable of? Even those who opposed the policies and actions of the United States—even those who thought the worst of us—acknowledged the genius and heroism of the *Apollo* program. With *Apollo*, the United States touched greatness.

When you pack your bags for a big trip, you never know what's in store for you. The *Apollo* astronauts on their way to and from the Moon photographed their home planet. It was a natural thing to do, but it had consequences that few foresaw. For the first time, the inhabitants of Earth could see their world from above—the whole Earth, the Earth in color, the Earth as an exquisite spinning white and blue ball set against the vast darkness of space. Those images helped awaken our slumbering planetary consciousness. They provide incontestable evidence that we all share the same vulnerable planet. They remind us of what is important and what is not. They were the harbingers of *Voyager*'s pale blue dot [refers to picture taken by *Voyager* outside Neptune's orbit that shows Earth as a "pale blue dot"].

We may have found that perspective just in time, just as our technology threatens the habitability of our world. Whatever the reason we first mustered the *Apollo* program, however mired it was in Cold War nationalism and the instruments of death, the inescapable recognition of the unity and fragility of the Earth is its clear and luminous dividend, the unexpected final gift of *Apollo*. What began in deadly competition has helped us to see that global cooperation is the essential precondition for our survival.

Travel is broadening.

It's time to hit the road again.

Columbia: The World's First Space Shuttle

By Ed Magnuson

Commander John Young flawlessly landed the space shuttle *Columbia* in California's Mohave Desert in April 1981. *Columbia* returned from space like an ordinary airplane, and unlike former spacecraft, it would be flown again. This first successful space shuttle trip happened at a time when Americans were suffering from a collective sense of futility and doom following the Vietnam War, and, consequently, the *Columbia*'s successful trip served as a tremendous boost to Americans' pride and spirit. Some said it was a much-needed reaffirmation of U.S. technological prowess and compared the fifty-four-and-a-half-hour shuttle trip to the first flight of Orville and Wilbur Wright at Kitty Hawk. Nations worldwide enthusiastically congratulated the United States for its achievement, although the Chinese expressed disapproval of both the United States and the USSR for casting "a shadow of war" over space exploration. Indeed, there were mixed reactions in the United States over the shuttle's potential military role, and some believed that the shuttle program could only be supported for military reasons.

In this selection, Ed Magnuson describes the landing of the *Columbia* space shuttle, the national and international responses to the event, and the scientific, commercial, and political significance of the space shuttle program. Ed Magnuson has worked for many years as a senior news writer for *Time* magazine.

S uddenly shouts rose from the hot, sunbaked desert floor in Southern California. There it was, high over the distant buttes, a tiny, gleaming dot in the pale blue sky, an apparition from space returning to earth.

Inside the cockpit, the 50-year-old commander, with glasses specially fitted into his helmet to correct the farsightedness of middle age, took over the controls for the final critical maneuvers. Expertly, the veteran pilot guided his craft through a long, easy turn. When he completed the maneuver, the ship was lined up perfectly with a runway marked in the ancient bed of Rogers Dry Lake six miles away. "Right on the money, right on the money!" encouraged Mission Control.

Then John Young edged the "stick" forward, and his ship's porpoise-shaped nose dropped slightly. Plunging earthward, *Columbia* was falling at an angle about seven times steeper than a normal airliner's descent and was traveling half again as fast. Powerful as it had been on takeoff, the ship was now functioning as a 102-ton glider with no engine to correct its course.

At 1,800 ft. and 35 sec. from landing, Young pulled back the stick to check his dive. Only *Columbia*'s stubby wings and slightly flared underbelly were giving it lift. But, to his delight, he found the craft far more aerodynamically buoyant than expected. Nineteen seconds before landing, he dropped his wheels.

"Gear down," reported a chase jet, buzzing alongside and counting off the attitude: "50 feet . . . 40 . . . 5—4—3—2—1—Touchdown!" As its rear wheels made contact, the flight director in far-off Houston told his tense crew: "Prepare for exhilaration." Nine seconds later, the nose wheels were down too. *Columbia* settled softly onto the lake bed. Young had floated the shuttle along 3,000 ft. beyond the planned landing spot, able to use its surprising lift to make a notably smooth touchdown. As it rolled to a stop through the shimmering desert air, The Star-Spangled Banner rattled forth from hundreds of portable radios tuned to a local station. From Mission Control in Houston's Johnson Space Center came an exuberant "Welcome home, *Columbia*. Beautiful. Beautiful."

So it was: simple and flawless, almost as if it had been performed countless times before. Yet the picture-perfect landing on California's Mojave Desert [in April 1981] all but obscured the historic nature of those last, breathtaking moments of *Columbia*'s 54.5-hr. odyssey. Gone were the great parachutes and swinging capsules of earlier space missions, splashing into the sea, never to travel into space again. For the first time, a man-made machine had returned from the heavens like an ordinary airplane—in fact, far more smoothly than many a commercial jet. So long delayed, so widely criticized, *Columbia*'s flight should finally put to rest any doubts that there will one day be regular commuter runs into the cosmos.

In the astonishing complexity of the craft's design, in its peerless performance, certainly in the cool performance of its astronauts-possessors of what Tom Wolfe calls "the right stuff"—*Columbia* was a much needed reaffirmation of U.S. technological prowess. It came at a moment when many Americans, and much of the world as well, were questioning that very capability. The doubts grew out of a succession of U.S. setbacks: from the defeat in Viet Nam to the downed rescue helicopters in the Iranian desert, from the debacle of Three Mile Island to Detroit's apparent defenselessness against the onslaught of Japanese cars. The flaming power of *Columbia*'s rockets seemed to lift Americans out of their collective sense of futility and gloom. At last they had a few things to cheer: an extraordinary spacecraft—the most daring flying machine ever built—and two brave and skilled men at its helm. As President Reagan told the astronauts, "Through you, we feel as giants once again."

Jubilant giants, at that. "The shuttle will become the DC-3 of space," exulted veteran Astronaut Deke Slayton, boss of orbital flight-test crews, referring to the sturdy Douglas aircraft that opened new routes for commercial aviation in the mid-1930s. *Columbia*'s maiden space voyage brought to mind the first flight of Orville and Wilbur Wright at Kitty Hawk, Lindbergh's lone-eagle crossing of the Atlantic, even the completion of the first transcontinental railroad in 1869, which would turn a land of remote frontiers into a nation. Princeton's prophet of space colonization, Physicist Gerard O'Neill, saw the flight as a first step toward establishing mining facilities on the moon. Still others spoke of the shuttle's potential role in scientific research, in space manufacturing, in the eventual tapping of solar energy in orbit, in controlling the new "high ground" of space against Soviet incursion.

From the instant of *Columbia*'s touchdown, a moment watched by tens of millions of television viewers in the U.S. and perhaps hundreds of millions more round the world, Americans seemed to go into orbit themselves. "Terrific!" shouted Dennis O'Connell, a truck driver from Queens, N.Y., as he paused in a Manhattan pub to watch the landing. "It shows everybody we're still No. 1." Mrs. Alicia Hoerter, a Louisville grandmother, could barely contain her excitement or her puns. "Go *Columbia*, the gem of a notion!" she exulted. "First, it's a rocket, then it's a spaceship, then it's a plane." In a packed Georgia Tech ballroom, great whoops of joy went up when John Young, class of '52, put *Columbia* down on the desert floor, and a band struck up "I'm a ramblin' wreck from Georgia Tech."

Not since the first landing of men on the moon had the nation shown such enthusiastic interest in space. Teachers interrupted classes so youngsters could see the landing. Work in offices and factories virtually ceased. Hearing that *Columbia* was about to touch down, a fitter in a Manhattan men's shop dashed off to the nearest TV set, leaving a customer standing before a mirror all pinned up in an unfinished suit. The *Atlanta Constitution*'s resident cartoonist, Baldy, showed a beaming Uncle Sam emerging out of the shuttle with his arms raised high like a victorious boxer's. Though some editorial writers expressed discomfort about the shuttle's military role, others dismissed such fears. Commented the *Chicago Tribune:* "It appears we will get into a space arms race whether we like it or not . . . So fly aloft, *Columbia*!; deliver your laser guns and satellite busters and spy eyes. Build your battlestars. May the Force be with us."

All but forgotten amid America's sudden love affair with the shuttle were its $9.9 billion price tag (at a 30% cost overrun), all those loose tiles, the exploding engines, even the last-minute computer failure, to say nothing of the inevitable jokes about America's "space lemon" and "flying brickyard." Could past scorn actually have increased the passion of this new embrace? The shuttle had become a kind of technological Rocky, the bum who perseveres to the end, the underdog who finally wins. *Columbia*'s success, explained Milwaukee Sociologist Wayne Youngquist, "ties in with so many of our cultural themes. It's Horatio Alger. It's The Little Engine That Could."

Perhaps. But the infatuation also had a boisterous, abrasive, decidedly chauvinistic tone. Out in the desert, many among the nearly one-quarter of a million people who had gathered to welcome the shuttle home sported T shirts emblazoned EAT YOUR HEARTS OUT, RUSSIANS. In a New York bar, after watching the landing, a patron boasted: "The French and the Brits can't do anything like that. Neither can the Russkis."

The French and the British, not to mention the Germans and Japanese, were not about to disagree. In London, the mass-circulation dailies exploded in a chorus of adulation. FANTASTIC! exclaimed the *Daily Mail.* WOW! trumpeted Rupert Murdoch's *Sun.* Most Britons, rather than showing concern over the shuttle's military potential, seemed to welcome it. Said the *London Times:* "The conquest of space is both a necessary expression of man's drive to explore and understand his environment and a military requirement if the West is not to be dominated by Soviet activity in space."

The West Germans had special reason to celebrate. They are the prime builders of Europe's main contribution to the shuttle program: the Spacelab, a self-contained scientific compartment for up to four experimenters scheduled to be carried aloft in 1983. Said one official: "Success for America means a breakthrough for us too and signals the entry of Western Europe into aerospace." The French, who are building a conventional rocket launcher called Ariane, which could draw away some of the shuttle's business, were no less effusive. Said *Le Figaro:* "After their political and military failures of recent years, our friends [the Americans] needed a big technological success. And they've got one." The French public wanted to share that success. During the very hour of *Columbia*'s homecoming, France's government-run television was to air a required, equal-time political broadcast for the April 26 presidential balloting. But viewers protested so vociferously that only twelve minutes before touchdown, France's election commission scrubbed the broadcast with the candidates' belated assent, and the French got to see le shuttle's return. "Reason," intoned *Le Figaro*, "triumphed at the last moment."

For Japanese televiewers, the landing occurred in the early hours before dawn, local time. But in a country that both admires and competes with American technology, some 2 million households tuned in for the event. In his message of congratulations to the U.S., Prime Minister Zenko Suzuki said of the shuttle: "It is the crystallization of your nation's highly developed technologies and scientific achievements and symbolizes the beginning of an 'American renewal.'"

The Soviets, again, complained that the shuttle is mainly a military vehicle. But they did show 30 seconds of the landing on televison. Chinese Communist newspapers, though fascinated by the idea of products labeled "made in space," excoriated both the U.S. and U.S.S.R. for casting a "shadow of war" over space.

Parked on the desert, *Columbia* had a decidedly unwarlike look. It survived its journey in remarkably fine style. A dozen or so of its 31,000 heat-shielding tiles had come unstuck during the thunderous ascent. But during its glowing, 2,700' plunge through the atmosphere, a maneuver that has been likened to riding inside a meteor, not one was lost from the craft's underbelly. Only a few tiles were gouged and chipped, apparently by pebbles and other desert debris kicked up by the wheels after touchdown. After an initial going-over at Edwards Air Force Base, the shuttle will be placed atop a modified Boeing 747 for a slow, two-day piggyback return to Cape Canaveral, where the ship will be refitted for a second launch, probably in September [1981]. The astronauts will be Joe Engle, 48, and Richard Truly, 43, this mission's back-up crew.

They can hardly outdo Young, who has now made five space flights, including a moon landing, and his rookie pilot, Bob Crippen, 43. Though their lift-off was delayed two days because of that computer failure, once they settled into the cockpit for the second try, everything went, well, like a rocket. Barely 45 min. off the launch pad, *Columbia* was circling the earth at an altitude of 150 miles. Before the end of the day it reached 170 miles. Meanwhile, two vessels steamed out to recover the 80-ton shells of two spent solid-fuel rockets that had parachuted into the Atlantic. When a nosey Soviet "trawler" edged into the site, the Coast Guard vessel *Steadfast* had to warn it off, then actually block its path, before the Russians backed off. The steel rocket frames were burned and bent a bit, but can probably be overhauled and refilled for another shot.

As always, there was in-flight banter between the astronauts and the Houston control center. When Crippen felt Houston was loading him with too many tasks at one time—realigning the inertial navigational unit, shooting a picture of the Southern Lights, confirming a message on the teletype—he asked in mock seriousness: "You mean all that right now?" To jog the astronauts awake, Houston piped in a loud country and western ditty about the shuttle called The Mean Machine. There was a somewhat more serious moment when Vice President Bush got on the radio from Washington to congratulate them on behalf of the nation.

There were also a few minor glitches. During the first "night" in space—actually they saw the sun rise and set once during every 90-min. orbit—Young and Crippen complained about a chill in the cabin. The temperature had dropped to 37 F. "I was ready to break out the long undies," joked one of the frozen astronauts. The problem was quickly fixed with a signal from earth that pumped warm water into the cabin's temperature control system. Young and Crippen had less luck fixing a faulty flight data recorder that had stopped mysteriously. They tried to get to it with a screwdriver but found the panel over it had been too tightly screwed down (or "torqued," as NASA put it).

The most serious problem came on the second night when an alarm light flashed and a bell jolted Young and Crippen out of their reveries. It was a warning of a malfunction in a heating unit on one of the three auxiliary power units for *Columbia*'s hydraulic systems, which control the landing gear and elevons [movable hinged sections of the wing]. The heater keeps the unit's fuel from freezing up. A throw of a switch got it working again, but *Columbia* is such a masterpiece of engineering redundancy that any one of the units could have saved the day. Said Flight Director Neil Hutchinson: "It's absolutely amazing. We didn't have anything that is a show stopper."

The real "show stopper," of course, might have been the landing. But it was breathtakingly "nominal," NASA lingo for "perfect." Crossing the coast below Big Sur at Mach 7, seven times the speed of sound, or about 5,100 m.p.h., Crippen crowed: "What a way to come to California!" Young lost his cool only after he had artfully landed *Columbia* right on the runway's center line. Eager to make an exit, he urged Houston to get the reception crews to speed up their "sniffing" chores—ridding the ship of noxious gases with exhausts and fans. When he was finally allowed to emerge, 63 min. after touchdown, he bounded down the stairs, checked out the tiles and landing gear, then jubilantly jabbed the air with his fists. It was probably Young's most uncontrolled move of the entire flight.

Curiously, Young's and Crippen's heartbeat patterns reversed on takeoff and landing. Both are normally in the 60s. At launch Young's rose only to 85 beats a minute, while Crippen's soared to 135. Returning, Young's pulse rate zipped up to 130 as he flew the craft in. Crippen's stayed around 85.

To be sure, Young's racing pulse slowed down soon after landing—and the nation's is likely to do the same. Says Forrest Berghorn, a political scientist at the University of Kansas: "The American spirit is too self-centered to concern itself with this for very long. The space shuttle success is in a class with our hockey victory over Russia." That may be too harsh a judgment; of late there have been signs of a renewed popular interest in space. Yet even those who want a redoubled U.S. space effort doubt there will be a lasting effect from the flight unless a profound change of mood occurs in budget-minded Washington. Says Jerry Grey, public policy administrator for the American Institute of Aeronautics and Astronautics: "Right now, there is no real commitment to space, no strong proponent of it within the Administration."

There is no doubt of continued military interest in the shuttle. But in the realms of pure science and commercial enterprise, the future of the costly space shuttle seems far from assured.

The Mir Space Station

By Clay Morgan

In this selection Clay Morgan explains the significance of Russia's *Mir* space station that orbited about 250 miles above Earth for fifteen years, traveling at an average speed of 17,885 miles per hour. Morgan says that the *Mir* station was the pride of Russia's early space progam, representing significant technological prowess and future potential for the country. The *Mir* space station was also a catalyst for improved international relations, especially between the United States and Russia, two countries which enjoyed their first significant technical partnership working together on the *Mir* project. *Mir* jointly hosted many American and Russian astronauts as well as numerous other international visitors. Some of the astronauts aboard *Mir* set records for long-duration space flights. Sergei Avdeyev, for example, spent a total of 747 days on the space station. Morgan explains that although the *Mir* station outlived its expected lifetime by ten years, the station eventually had structural problems that led Russia's prime minister Mikhail Kasyanov to "deorbit" the aging station in 2001.

Clay Morgan is an author, columnist, and essayist who has written several publications for NASA. He teaches writing at the University of Houston, Texas, where he started the first master's program in space exploration and study. Morgan is a former commentator for National Public Radio's *All Things Considered.*

T he Russian Space Station Mir endured 15 years in orbit, three times its planned lifetime. It outlasted the Soviet Union that launched it into space. It hosted scores of crewmembers and international visitors. It raised the first crop of wheat to be grown from seed to seed in outer space. It was the scene of joyous re-

Clay Morgan, *Shuttle-Mir (Mir-Shattl): The United States and Russia Share History's Highest Stage.* Houston, TX: National Aeronautics and Space Administration, 2001.

unions, feats of courage, moments of panic, and months of grim determination. It suffered dangerous fires, a nearly catastrophic collision, and darkened periods of out-of-control tumbling.

Mir soared as a symbol of Russia's past space glories and her potential future as a leader in space. And, it served as the stage—history's highest stage—for the first large-scale, technical partnership between Russia and the United States after a half-century of mutual antagonism.

Mir did all of that and, like most legends, was controversial and paradoxical. At different times and by different people, Mir was called "venerable" and "derelict." It was also "robust," "accident-prone," and "a marvel," as well as "a lemon."

What "Mir" Means

For Russians, the very name "Mir" held meaning, feeling, and history. Mir translates into English as "world," "peace," and "village"; but a single-word translation misses its full significance. Historically, after the Edict of Emancipation in 1861, the word "mir" referred to a Russian peasant community that owned its own land. A system of state-owned collective farms replaced the mir after the Russian revolution of 1917.

As with most legends, Mir was literally beyond the reach of most men and women; but it could be seen by many as a bright light arcing across the night sky. Mir undoubtedly provoked many thoughts around the globe about who we—as a human race—are and where we are going.

The cosmonauts and astronauts who were fortunate enough to travel to Mir were always impressed by its appearance. Regardless, Mir remained difficult to describe. Someone once called Mir a 100-ton Tinker Toy®, a term that recalled Mir's construction. Adding modules over the years, and then sometimes rearranging them, the Russians had built the strangest, biggest structure ever seen in outer space. Traveling at an average speed of 17,885 mph, the space station orbited about 250 miles above the Earth. Mir was both great and graceful—and incongruous and awkward—all at the same time.

In outward appearance, Mir has also been compared to a dragonfly with its wings outstretched, and to a hedgehog whose spines could pierce a spacewalker's suit. NASA-4 Astronaut Jerry Linenger compared Mir to "six school buses all hooked together. It was as if four of the buses were driven into a four-way intersection at the same time. They collided and became attached. All at right angles to each other, these four buses made up the four Mir science modules. . . . Priroda and Spektr were relatively new additions . . . and looked it—each sporting shiny gold foil, bleached-white solar blankets, and unmarred thruster pods. Kvant-2 and Kristall . . . showed their age. Solar blankets were yellowed . . . and looked as drab as a Moscow winter and were pockmarked with raggedy holes, the result of losing battles with micrometeorite and debris strikes over the years."

On the Inside

On the inside, Mir often surprised people, too, even when they thought they were ready for the view. By the time Americans arrived on Mir—nearly a decade into its life—the station had become cluttered with used-up and broken equipment and floating bags of trash. During Mir's lifetime, no adequate remedy was ever developed to deal with the stowage situation. Mir looked like a metal rabbit warren, or, as Mike Foale [an American astronaut who visited Mir] put it, "a bit like a frat house, but more organized and better looked after."

Still, Mir was home and shelter to its crews, and how it looked to them depended on their perspectives and situations. The ivory-like controls of the Base Block [the core module] reminded [American astronaut] David Wolf of classic science-fiction stories, such as *The Time Machine* by H.G. Wells. After a fly-around in the cramped Soyuz capsule, Linenger wrote: "Looking into the station I could see a lone ray of light shining through the port window and outlining the dining table. We had left some food out for dinner. It was the only time during my stay in space that Mir looked warm, inviting, and spacious. It reminded me of opening the door to a summer cottage that had been boarded up for the winter, looking inside, and seeing familiar surroundings."

Mir Set Records

Mir set every record in long-duration spaceflight. Physician Valeri Polyakov lived aboard Mir for a single, continuous-orbit stay of 437 days, 17 hours, and 38 minutes. He completed his stay in 1995 as American Norm Thagard began his Mir residency. Polyakov's experiences contributed greatly to the biomedical studies of long-term human spaceflight conducted by the Institute of Biomedical Problems, where he served as Deputy Director. Combined with an earlier Mir expedition flight, the Russian cosmonaut spent a total 678 days, 16 hours, and 33 minutes on the Russian space station. However, his achievement for total time in space was surpassed in 1999 by Sergei Avdeyev, who endured a total 747 days, 14 hours, and 12 minutes during three space missions. During Shuttle-Mir, Shannon Lucid set the space endurance record for women in 1996 when she spent 188 days, 4 hours, and 00 minutes in orbit.

Just as "mir"—the word—had many meanings for Russians, Mir—the place—provoked many different feelings. In February 1995, Russian Cosmonaut Vladimir Titov flew aboard the "near Mir" flight, STS-63, when the Shuttle rendezvoused with Mir. Six years earlier, Titov had spent a year aboard Mir as an expedition member, when Mir consisted of only the Base Block, the two Kvant modules, a Soyuz, and a Progress spacecraft. About seeing Mir again, Titov said, "It was very wonderful, a wonderful view." STS-63 did not dock, but Titov visited Mir again as a crewmember of STS-86.

Mir Is Deorbited

Alas, the sturdy Mir was built on a sinking foundation. Without repeated boostings, all things in low Earth orbit must eventually come down. With the new International Space Station requiring much of the Russian space program's attention and financing, the Mir space station was doomed to be deorbited. A strong effort rallied in Russia to keep Mir aloft; and, at one point, Russian State Duma representatives were calling for the firing of Yuri Koptev from his post as the head of Russia's aerospace agency. However, on December 30, 2000, Russian Prime Minister Mikhail Kasyanov signed a resolution calling for Mir to be sunk into the ocean early in 2001 [which it was].

The Space Shuttle Challenger Disaster

By Nick Greene

Many remember where they were and what they were doing in 1986 when they saw or heard about the space shuttle *Challenger* explosion. In this selection Nick Greene recounts some of the details of the *Challenger* accident. Greene explains that the *Challenger* had flown nine successful missions, but that the 1986 mission was unique because it included the TISP, Teacher in Space Program. Selected among eleven thousand applicants, Sharon Christa McAuliffe was chosen to be the first teacher to fly into space.

Greene explains that after numerous problems and many delayed liftoffs, the *Challenger* was finally launched on the morning of January 27, 1986. Seventy-three seconds after liftoff, he says, the *Challenger* exploded, killing all seven crew members. A stunned nation had witnessed the first American astronauts to be killed in flight. Following the disaster, the entire shuttle program was grounded for investigations into the accident. Officials found that not only had a failed O-ring seal been the problem, but that engineers had been concerned about the proper functioning of the O-ring seal before the shuttle's launch and NASA officials had gone forward with the *Challenger* mission nonetheless. As a result, NASA management implemented stricter regulations over quality control and safety of the space shuttle program, which did not launch another shuttle until 1988.

Families of the *Challenger* crew formed *Challenger* organizations comprised of forty-two learning centers around the United States, in Canada, and in the United Kingdom. The centers provide space-related educational resources for students, teachers, and parents.

Nick Greene is an electronics technician, software engineer, and

Web page designer. Greene is an amateur astronomer, a member of the Planetary Society, cofounder of the Antarctic Astronomy Society, and UN World Space Week coordinator for the continent of Antarctica.

I t was a NASA tragedy.
NASA's Shuttle program was begun in the 1970s, to create reusable craft for transporting cargo into space. Previous space craft could only be used once, then were discarded. The first shuttle, *Columbia,* was launched in 1981. One year later, the Shuttle *Challenger* rolled off the assembly line as the second shuttle of the US fleet. They were followed by the shuttles *Discovery* in 1983 and *Atlantis* in 1985.

The Space Shuttle *Challenger* flew nine successful missions before that fateful day of the disaster in 1986.

The Challenger's Mission

Shuttle mission 51L was much like most other shuttle missions. The *Challenger* was scheduled to carry some cargo, the Tracking Data Relay Satellite-2 (TDRS-2), as well as fly the Shuttle-Pointed Tool for Astronomy (SPARTAN-203)/Halley's Comet Experiment Deployable, a free-flying module designed to observe tail and coma of Halley's comet with two ultraviolet spectrometers and two cameras.

One thing made this mission unique. It was scheduled to be the first flight of a new program called TISP, the Teacher in Space Program. The *Challenger* was scheduled to carry Sharon Christa McAuliffe, the first teacher to fly in space.

Selected from among more than 11,000 applicants from the education profession for entrance into the astronaut ranks, McAuliffe was very excited about the opportunity to participate in the space program. "I watched the Space Age being born and I would like to participate."

Besides McAuliffe, the *Challenger* crew consisted of mission commander Francis R. Scobee; pilot Michael J. Smith; mission specialists Ronald E. McNair, Ellison S. Onizuka, and Judith A. Resnik; and payload specialist Gregory B. Jarvis. Christa was also listed as a payload specialist.

Plagued by Problems from the Start

From the beginning, though, Shuttle Mission 51L was plagued by problems. Liftoff was initally scheduled at 3:43 P.M. EST on January 22, 1986. It slipped to Jan. 23, then Jan. 24, due to delays in mission 61-C and finally reset for Jan. 25 because of bad weather at transoceanic abort landing (TAL) site in Dakar, Senegal. The launch was again postponed for one day when launch processing was unable to meet new morning liftoff time. Predicted bad weather at Kennedy Space Center (KSC) caused the launch to be rescheduled for 9:37 A.M. EST, Jan. 27, but it was delayed another 24 hours when a ground servicing equipment hatch closing fixture could not be removed from the orbiter hatch.

The fixture was sawed off and an attaching bolt drilled out before closeout completed. During this delay, the cross winds exceeded limits at KSC's Shuttle Landing Facility. There as a final delay of two hours when a hardware interface module in the launch processing system, which monitors the fire detection system, failed during liquid hydrogen tanking procedures. The Challenger finally lifted off at 11:38:00 A.M. EST.

Seventy-three seconds into the mission, the *Challenger* exploded, killing the entire crew.

Seventy-three seconds after liftoff, the Challenger *exploded, killing all seven crew members, including schoolteacher Christa McAuliffe (far left).*

The Aftermath

The reaction was immediate, from the crowds of family and friends gathered to watch the launch of the Space Shuttle *Challenger* to the millions tuned in worldwide, most people were stunned. In a speech later that day, President Ronald Reagan expressed the feelings of many who were grieving.

"Today is a day for mourning and remembering," he said. "Nancy and I are pained to the core over the tragedy of the shuttle *Challenger.* We know we share this pain with all of the people of our country. This is truly a national loss. Nineteen years ago almost to the day, we lost three astronauts in a terrible accident on the ground. But we've never lost an astronaut in flight. We've never had a tragedy like this. And, perhaps, we've forgotten the courage it took for the crew of the shuttle. But the Challenger Seven were aware of the dangers and overcame them and did their job brilliantly. We mourn seven heroes."

Afterwards, a special commission to investigate the cause of the Space Shuttle *Challenger* accident was appointed by President Reagan. Headed by former secretary of state William Rogers the commission included former astronaut Neil Armstrong and former test pilot Chuck Yeager.

The Failure of an "O-ring" Seal

The commission's report cited the cause of the disaster as the failure of an "O-ring" seal in the solid-fuel rocket on the Space Shuttle *Challenger*'s right side. The faulty design of the seal coupled with the unusually cold weather, let hot gases leak through the joint. Booster rocket flames were able to pass through the failed seal enlarging the small hole. These flames then burned through the Space Shuttle *Challenger*'s external fuel tank and through one of the supports that attached the booster to the side of the tank. That booster broke loose and collided with the tank, piercing the tank's side. Liquid hydrogen and liquid oxygen fuels from the tank and booster mixed and ignited, causing the Space Shuttle *Challenger* to tear apart.

The commission not only found fault with a failed sealant ring but also with the officials at the National Aeronautics and Space Administration (NASA) who allowed the shuttle launch to take place despite concerns voiced by NASA engineers.

The entire space shuttle program was grounded during the Space Shuttle *Challenger* Commission's investigation and did not resume flying until shuttle designers made several technical modifications and NASA management implemented stricter regulations regarding quality control and safety. Shuttle missions resumed on September 28, 1988, with the flight of the shuttle *Discovery*.

Remembering the Crew

In 1991, the shuttle *Endeavour* joined the fleet to replace the *Challenger*, again bringing the number of ships to four. Over the next few months and years, the family members of the crew of shuttle *Challenger* flight 51L dealt with their grief. With the support of family and friends, as well as people worldwide who joined in the grieving process, they lived day to day, and though they will never forget, they managed to continue with their lives.

As a tribute to the memories of their loved ones, the families helped form the *Challenger* Organization, which provide resources for students, teachers, and parents for educational purposes. Included in these resources are 42 Learning Centers in 26 states, Canada, and the UK which offer a two-room simulator, consisting of a space station, complete with communications, medical, life, and computer science equipment, and a mission control room patterned after NASA's Johnson Space Center and a space lab ready for exploration.

For many, the *Challenger* disaster was a kind of life-altering event, as the assassination of President [John F.] Kennedy was for an older generation. Many of us will always remember where we were and what we were doing when we witnessed or heard about the explosion. Most of us consider the lost crewmembers to be heroes.

The evening of the explosion, President Reagan said it well, "The crew of the space shuttle *Challenger* honored us in the way in which they lived their lives. We'll never forget them nor the last time we saw them this morning as they prepared for their journey and waved goodbye and slipped the surly bounds of Earth and touched the face of God."

Learning from the Columbia Space Shuttle Accident

By Sean O'Keefe

In his testimony before the House Science Committee in September 2003, NASA administrator Sean O'Keefe made his final report on the *Columbia* space shuttle accident. On February 1, 2003, the space shuttle *Columbia* disintegrated just a few minutes before its scheduled landing in Florida, destroying the craft and killing the seven crew members on board. As NASA administrator, O'Keefe assured the committee that NASA would follow the safety measures that the Columbia Accident Investigation Board (CAIB) had recommended. Indeed, O'Keefe said that before the CAIB had made its recommendations, NASA had pursued its own agency-wide efforts to improve the space shuttle program and had established a new NASA Engineering and Safety Center. O'Keefe said that to comply with the CAIB recommendations, NASA had created a Return to Flight Implementation Plan that improved technical, organizational, and managerial aspects of NASA. The NASA administrator assured the committee that the space program would continue and that he believed a new chapter in NASA history was beginning; one more committed to excellence and safety.

Sean O'Keefe was appointed by President George W. Bush as the tenth administrator of NASA in December 2001. O'Keefe has held many positions including deputy director of the Office of Management and Budget, professor of business and government policy at Syracuse University, New York, where he was also director of national security studies. O'Keefe served as the secretary of the navy under President George H.W. Bush in 1992. He served as comptroller and chief financial officer of the Department of Defense and was

Sean O'Keefe, testimony before the House Science Committee, Washington, DC, September 10, 2003.

staff director of the Defense Appropriations Subcommittee. O'Keefe has written several articles and coauthored the book *The Defense Industry in the Post–Cold War Era.*

Shortly after the tragic loss of the Space Shuttle and its heroic crew, I made a solemn pledge to the families of *Columbia*'s crew that we would find out what caused the loss of the Space Shuttle *Columbia* and its crew, correct what problems we find, and safely continue with the important work in space that motivates our astronauts and inspires millions throughout the world. Thanks to the CAIB's (Columbia Accident Investigation Board) thorough report, we now definitively know what caused the accident. It was a combination of hardware, process and human failures. We also have a more complete understanding of the problems that must be fixed at NASA to ensure that Space Shuttle operations are conducted as safely as humanly possible in pursuit of our Nation's space exploration and research agenda.

The CAIB report provides NASA with a very detailed roadmap for returning to flight safely, one that we intend to faithfully follow. I can assure you that we will not only implement the CAIB's recommendations to the best of our ability, but we are also seeking ways to go beyond their recommendations.

Today's focus is on the hard lessons we've learned from the *Columbia* accident and about the hard work that lies ahead before we are ready to launch the Space Shuttle *Atlantis* for the STS-114 mission. I want to emphasize, as we undertake this work, we will be ever mindful of and appreciative of the people who have helped NASA and our entire country recover from that terrible first day of February.

Gratitude to the Families of the Columbia Crew

First and foremost, we owe enormous gratitude to the brave families of the *Columbia* crew. Through their steadfast courage and dignity they have provided inspiration to the Nation. A fitting memorial for the crew will be constructed at Arlington National Cemetery. We thank the members of this Committee for your strong support of the Columbia Orbiter Memorial Act, which President [George W.] Bush signed into law on April 16, 2003.

[In August 2003] the family members demonstrated an incredible spirit of exploration and discovery in their own right as they joined astronaut Scott Parazynski in climbing to the top of the recently named Columbia Point, a prominent vista on Colorado's Kit Carson Mountain that now honors the memory of the *Columbia* STS-107 crew.

Recovery of Space Shuttle Parts

We are also indebted to the over 14,000 people from the Environmental Protection Agency, Federal Emergency Management Agency, the Federal Bureau of Investigation, Defense Department, U.S. Forest Service, the Texas and Louisiana National Guards and many state and local law enforcement and emergency service units who contributed to the recovery of *Columbia*'s debris. As a result of this unprecedented interagency and intergovernmental cooperative effort, an area in eastern Texas and western Louisiana, about the size of Rhode Island, was carefully searched, resulting in the recovery of thirty-eight percent of the dry weight of the orbiter, including several key parts from the left wing, the part of the Orbiter damaged by a foam strike during liftoff, and the critical Orbiter Experimental Recorder—the data recorder that verified and validated much of what was learned about the accident. We are deeply saddened to note that one of the helicopters searching for debris from the Shuttle *Columbia* crashed in the Angelina National Forest in east Texas on March 27, claiming the lives of the pilot and a Forest Service Ranger. Our thoughts and prayers go out to the families of the helicopter crew members.

In support of this unprecedented operation, we received tremendous hospitality and support from the Texas communities of Lufkin, Hemphill, Nagadoches, Palestine and Corsicana, as well as the Louisiana communities of Shreveport and Leesville, particularly in support of activities at Barksdale (AFB) [Air Force Base] and Fort Polk. NASA vows not to forget the many kindnesses bestowed upon our people and the other recovery workers by all these communities. We will use the resources and people of our Education Enterprise to help nurture the spirit of discovery and exploration in the young people who grow up in the region, just as we are working to help inspire and motivate school children throughout the country as they embark on their studies this fall.

Finally, we are grateful for the diligent work of the Columbia Accident Investigation Board members and staff. As many of you know, the Board has worked non-stop since it was given this important responsibility. [Chairman] Admiral Gehman has performed many tremendous acts of public service throughout his distinguished career, and I'm certain that the history books will regard his work on this report as among his most significant contributions to his country.

Return to Flight Implementation Plan

We accept the findings of the Board and will comply with its recommendations. The Columbia Accident Investigation Board's report recommendations will be our benchmark for Return to Flight. Using the Board's recommendations as NASA's organizing principles for emerging from the *Columbia* accident as a safer, stronger and smarter organization, we have developed a preliminary Return to Flight Implementation Plan which details the Agency's evolving blueprint for returning to flight safely and reliably. Released on September 8 [2003], this preliminary Implementation Plan provides an outline of how NASA will comply with the recommendations of the Columbia Accident Investigation Board, and also includes other corrective actions. The Implementation Plan is a living document and will be updated on a regular and frequent basis, with input from across the entire Agency.

Following the logic of the Board's report, the preliminary Implementation Plan focuses on making improvements in the following key areas:

Technical excellence—Making specific technical engineering changes that will enhance our overall technical capabilities. Among these changes is the establishment of our new NASA Engineering and Safety Center at the Langley Research Center in Hampton, Virginia that will draw upon talent throughout our Agency to take a no holds barred approach to mission safety. If people in the center spot a problem or potential problem during their engineering and safety assessments of all our programs, they will be empowered to get the entire Agency, if necessary, focused on finding and implementing solutions.

Management—Putting in place more effective management procedures, safeguards, and decision-making processes.

* Organizational Culture—NASA recognizes that prior to the

Columbia mission, cultural traits and organizational practices within the Agency detrimental to safety were allowed to develop. We will now work diligently to develop an organizational culture that reflects the best characteristics of a learning organization, one based on clear and open communications throughout our Mission Teams, with a management culture that empowers both dialogue and achievement.

Further Improvements to the Space Shuttle Program

At the same time the CAIB was developing its report, NASA pursued an intensive, Agency-wide effort to identify additional actions that will further improve the Space Shuttle Program. We took a fresh look at all aspects of the Program, from technical requirements to management processes, and developed a set of internally-generated actions that complement and go well beyond the CAIB recommendations. For example, some of the types of activities we are focusing on include rudder speed brake actuator inspections and re-evaluation of catastrophic hazard analysis, to name a few.

The Implementation Plan integrates the CAIB recommendations as well as other actions. It is the first installment in a living document that will be periodically updated to reflect the progress toward safe return-to-flight and faithful implementation of the CAIB recommendations.

Anticipated Costs

With respect to preliminary budget implications of the return-to-flight efforts, on September 4, 2003, NASA submitted to the Committee an update to the FY [Fiscal Year] 2003 Operating Plan. This update reflects anticipated costs of about $40 million associated with implementation of an initial set of actions tied to the CAIB recommendations and other corrective actions. NASA is determining the full spectrum of recommended return-to-flight hardware and process changes, as well as their associated costs. The Administration is also assessing the long-term implications of the return-to-flight requirements. We will keep the Committee informed as decisions are made.

We are now determined to move forward with a careful, milestone-driven return to spaceflight activities, and to do so with the utmost concern for safety, incorporating all the lessons learned from the tragic events of February 1st. That's exactly what we will do.

Spaceflight Professionals Will Lead the Way

Our Return to Flight effort involves a team of spaceflight professionals, led at NASA headquarters by Dr. Michael Greenfield, Associate Deputy Administrator for Technical Programs and veteran astronaut Bill Readdy, Associate Administrator for Space Flight.

Another veteran astronaut, Jim Halsell, who has flown on five Shuttle missions, will oversee the day-to-day work required for our return to flight. As the commander of an upcoming shuttle mission, STS-120, Jim has a personal interest in ensuring that Return to Flight is done right. I can assure you we will also rely on the advice and judgment of all members of the astronaut corps, the men and women who have the most vested interest in safe operations of the Shuttle program.

We will also have the benefit of the wisdom and guidance of a seasoned Return to Flight Task Group, led by two veteran astronauts, Apollo commander Thomas Stafford and Space Shuttle commander Richard Covey. Members of the Stafford-Covey Task Group were chosen from among leading industry, academia and government experts. The Members of the Task Group have knowledge and expertise in fields relevant to safety and space flight, as well as experience in leadership and management of complex programs. The diverse membership of the Task Group will carefully evaluate and publicly report on the progress of our response to implement the CAIB's recommendations.

There is another body that NASA will greatly rely on in the Return to Flight process: this committee, and all in Congress who have a vital interest in how NASA performs our work on behalf of the American public. We very much respect and value this Committee's oversight responsibility, and I personally look forward to working with the Committee in the weeks and months ahead to ensure that we do our job right.

A New Chapter in NASA History

Building upon work already underway to address issues previously identified by the CAIB, the release of our preliminary Implementation Plan marks an important step in our efforts to address and fix the problems that led to the *Columbia* accident. We are about to begin a new chapter in NASA history, one that will be marked by a renewed commitment to excellence in all aspects of our work, a strengthening of a safety ethos throughout our organization and an enhancement of our technical capabilities.

As we proceed along this path, all of us will be challenged. I am absolutely certain that the dedicated men and women of NASA are up to this challenge and we will not let the families of the *Columbia* astronauts and the American people down.

The Impact of the Columbia Accident on the International Space Station

I would also like to provide an update on the status of the International Space Station (ISS) and the impact from grounding the Space Shuttle. The Space Shuttle's return to flight is critical to complete assembly and ensure research capability for the ISS. Only the Shuttle can deliver the large elements, spare parts and the logistics required to successfully meet our research goals and international agreements. While the Space Shuttle fleet is grounded as a result of the *Columbia* accident, Russian Soyuz and Progress vehicles continue to provide assured crew and cargo access to and from the ISS.

In the absence of Space Shuttle support, NASA and the International Partners [those countries involved in the ISS] are addressing contingency requirements for the ISS for the near- and long-term. In order to keep the *Expedition 7* and future crews safe, we are ensuring that there are sufficient consumables, that the ISS can support the crew, and that there is a method for safe crew return available.

The ISS *Expedition 7* crew (Yuri Malenchenko and Ed Lu) continue their stay on-board the ISS, which began in late April 2003. The ISS was re-supplied with a Progress vehicle (ISS Fight 12P) launched on August 28 and docked to the Station on August 30, 2003. The crew is continuing experiments for which sufficient hardware and supplies are already on board the ISS. Twenty-six science investigations are in process or planned for Increments [Expeditions] 7 and 8. Operations continue to go well, with sufficient consumables on board the ISS. The launch of the next Progress to resupply the ISS has been accelerated from January 2004 to November 2003. I am proud that the ISS partnership has come to-gether as a true partnership during this challenging time. I also wish to assure you that there is no schedule pressure to return the Space Shuttle to flight until we are confident it is safe to fly.

The *Expedition 8* crew (Commander C. Michael Foale and Flight Engineer Alexander Kaleri) is scheduled to accept handover of the ISS from the *Expedition 7* crew following their launch on Soyuz in October, 2003.

The U.S. Journey into Space Will Continue

In closing, I want to reiterate that the country owes Admiral Gehman and the entire Board a tremendous debt of gratitude for the service it has performed. We embrace the CAIB report and we are committed to implementing the recommendations and safely returning to flight.

Finally, I believe it is important to note that all 13 CAIB members arrived at and agreed to the final conclusion of their report: "The United States should continue with a Human Space Flight Program consistent with the resolve voiced by President George W. Bush on February 1, 2003: 'Mankind is led into darkness beyond our world by the inspiration of discovery and the longing to understand. Our journey into space will go on.'"

CHAPTER 3

Unmanned Space Exploration

The First Spacecraft to Search for Life on Mars

By U.S. News & World Report

The following selection is an interview with former internationally recognized planetary scientist Carl Sagan in 1976, the year the *Viking I* spacecraft made the first successful landing on Mars. Sagan, an expert on martian science, explains that scientists' excitement over pictures and information sent from the *Viking I* was due to the discovery of life-supporting elements such as nitrogen and argon in the martian atmosphere. The *Viking I* also took pictures of the planet's surface, revealing colossal landslides and river channels, phenomena also associated with the existence of life. Sagan suggests that these findings broaden the possibility of life on other planets, in other solar systems, and that humans might one day communicate with these life-forms. The scientist stresses the importance of humans following their natural curiosity, intelligence, and sense of adventure because doing so allows them to see what nature has done on other planets, giving profound insights into how Earth works.

U.S. News & World Report is a weekly newsmagazine that provides world and international news. It covers current events on business and economic issues, as well as practical news on personal finance, health, and education.

Q: Professor Sagan, why are you planetary scientists so ecstatic about the findings thus far of the Viking mission to Mars?

A: Because there are so many areas in which we have acquired fundamental new information. We have found nitrogen and argon in the Martian atmosphere. Nitrogen is a life-supporting element that is widely considered essential to any theory that life exists on Mars. Argon is a marker of a denser early atmosphere which many scientists feel was required for the origin of life.

There are more than a thousand exquisite photographs of Mars taken by Viking 1 in orbit which reveal a range of new processes, or ones previously only suspected. There are landslides on a colossal scale, and a set of tributary and major river channels. This seems quite good evidence that there was extensive rainfall in the ancient Martian past, indicating that there were more clement conditions favorable to life some hundreds of millions of years ago.

The photos the Viking lander is taking now that it is on the Mars surface are just as exciting. We see rolling hills and a rock-strewn landscape on one side, beautiful sand dunes on the other. We have a pinkish sky which is due to particles suspended in the atmosphere. The initial results of the microbiology instruments are tantalizing.

In sum, this amazingly rich set of data from a landing site that was chosen for its blandness makes one wonder if there aren't far more exciting areas on Mars.

Life on Mars

Q: What difference will it make to us here on Earth if there is life on Mars?

A: There are all sorts of answers to that question. One important answer is that all life on the Earth is the same, despite some external differences in form. In the essential biochemistry, every one of us—from bacterium to human being—is composed of the same proteins and the same nucleic acids, all put together in the same way. Therefore, the biologists have no idea what is possible in living systems—what ranges of biologies can exist. We have only one example.

Q: Could it be that, perhaps, over those rolling hills on Mars there may be some Earthlike people?

A: No, just exactly the opposite. Life on the Earth is all the same, and the reason for that is that we are all descended from a single instance of the origin of life. We're similar because we're related—we bacteria and we people.

On another planet, the statistical factors work differently, the physical environment is different, evolution follows other pathways, and one would expect that the organisms, if any, would be astonishingly different—not just in their external appearance, but more fundamentally in their internal make-up. The discovery of life on Mars—even very simple life—would work a fundamental revolution in biology which would have a wide range of practical implications for us here on Earth.

If we find life on Mars, we will have looked at two planets and found life on both of them—the Earth and Mars. That would go a long way toward arguing for the grand conclusion of an inhabited cosmos. If so, it seems very likely that on many planets in other solar systems life would have evolved into advanced forms. The conclusion naturally presents itself that we might be able to communicate, using radio telescopes, with intelligent beings on planets of other stars. That's why it will be so exciting if we do find life on Mars. On the other hand, if we do not find life on Mars, that would hardly foreclose the possibility of life on other planets.

Q: When Viking 2 lands in September, do you expect it to yield findings different from those at the site of Viking 1?

A: The primary reason we're going to these high northern latitudes—about 44 north on Mars—is because there is such an expectation. The amount of water vapor in the atmosphere there is larger. The atmospheric pressures are greater; therefore, the chance of at least interstitial liquid water in the Martian soil is larger. And since the Viking biology experiments are strongly oriented toward looking for organisms which enjoy liquid water, as they do on the Earth, that would improve the chance of finding life with the Viking instruments.

In addition, it's a very different geological region, and I suspect the pictures from there may be even more exciting than those we've seen from the Viking 1 landing site.

Water on Mars

Q: When you speak of "interstitial water," what do you mean?

A: Small quantities of a liquid-water film which might coat subsurface particles, at least for a brief period each day.

Q: Like dew?

A: Something like dew. It doesn't sound like much to drink for us, but it would be a lake for a microbe.

Q: You've suggested that Mars once had rivers, even full-flowing rivers—

A: That's right. In fact, I think that's very hard to gainsay right now.

Q: Where did these rivers go? What happened to them?

A: In order to have liquid water on Mars, you have to have a greater atmospheric pressure than is there today. It's as if you had a pan of water and carried it into a vacuum. It would all boil away very rapidly. You need to keep a lid on the liquid water by having a high-enough atmospheric pressure. Today we do not have that atmospheric pressure, and a pan of water exposed to the Martian environment would quickly boil and evaporate away. Therefore, if there was liquid water in the past, as the riverlike channels certainly seem to indicate, there must have been a higher pressure.

You ask: Where is all that water now? And the answer is surely not that it has escaped to space, because only a negligible amount of water can escape to space from Mars. So it has to still be there, but not in the liquid phase. There are a number of obvious choices. One is that the water is locked away in the polar caps, which we know are very recent deep deposits of some frozen volatile substance. Or the water could be locked away in the subsurface as permafrost, or chemically combined with the surface to produce hydrated minerals. We have some direct spectroscopic evidence that such is the case.

So, in effect, Mars has plenty of water today; it's just not in the liquid phase. And the denser ancient atmosphere seems to be buried away as well.

More-Advanced Organisms

Q: Is it possible Mars might belong to somebody else?

A: Yes. If there are microbes on Mars, I would consider that in some sense the planet belongs to them. And I don't consider it out of the question that there are more-advanced organisms waiting to be discovered.

Q: What sort of environment would they be living in?

A: We have a weather station on Mars. To answer your question, let me give you a typical weather report. It would be something like this: In the Chryse Basin today, the wind was from the northwest at 12 miles an hour, changing at midnight to from the south at 19 miles an hour. The temperature just before dawn was –122 Fahrenheit, rising to –20 Fahrenheit at 2 P.M. The barometric pressure is constant at 7.7 millibars. Chances for rain: zero.

Q: What about those high winds that we've heard about?

A: We've purposely landed at a time and place of very low winds because they could cause substantial damage, especially during the landing maneuvers. If, as is likely, the Viking lander survives for six months or a year, we should see substantially higher winds. Velocities up to 200 miles an hour are not infrequent over a Martian year, and even higher winds may be possible.

Manned Missions to Mars

Q: Does the success of Viking enhance the prospects of someday sending manned flights to Mars?

A: I'm not sure. One important point is that a manned mission to Mars would probably cost hundreds of billions of dollars. That seems, at least to me, an unconscionably large amount of money.

If we wished to make a crash effort, the technology is—or will shortly be—available. But why would we want to do it? We can't do everything that we're capable of; we have only finite resources. The clear lesson of Viking is that intelligent machines are able to do superbly well in exotic and at least partially hostile environments.

I think that the principal implication about future exploration of Mars is to go back with a better Viking—an advanced roving vehicle. You look at those pictures—especially in stereo—and you have this powerful emotion: You'd love to look over the next hill. In fact, in a few places the nearby hills dip down and you can see a little bit of the adjacent valley. You wonder what's there, and wish you could go and take a look.

Viking 1 is immobile. It can't move even an inch. But it's possible to put tractor treads on a Viking-type spacecraft, land it on Mars and have it wander. During a reasonable mission lifetime, it could cover many hundreds of miles.

Q: Do you feel we should continue to explore Mars rather than shifting attention to Venus or the moons of Jupiter?

A: I think we must have a balanced strategy. For all we know, the most exciting place in the solar system might be, say, Titan, the big moon of Saturn. If we never do reconnaissance missions out there, we'll never have an idea whether that is the case.

It's perfectly feasible, since these unmanned missions are relatively inexpensive. What I would recommend is intensive exploration of Mars by roving vehicles along with reconnaissance of much of the rest of the solar system, certainly including Jupiter and Saturn and their moons.

Q: What kind of money are we talking about to finance that?

A: Viking is by far the most expensive unmanned mission that the United States has ever flown, and it cost under a billion dollars. A single Viking rover would cost something like a third of that, because many components already exist.

The typical cost of any of these explorations is in the low hundreds of millions for a one-or-two-spacecraft mission. A decade of full-scale exploration of the solar system would cost about the same as a couple of Trident submarines.

A Unique Opportunity

Q: What return would the citizens of the U.S. get from this expenditure?

A: I think civilizations are known in the long perspective by what they do—by the historical turning points which they initiate. We now have, because of our technology, a unique opportunity to explore our surroundings in space.

The practical advantages are immense, even in the medium-range time frame. The return does not come the year after you go there; it takes a little time. But the opportunity to compare our planet with other planets that have had different histories gives us a perspective on the alternative fates of worlds.

We live on a world that is structured in a certain way, and we wish to understand how to make it go in one direction rather than another. You can't experiment on it, because if you make a mistake you're in trouble—consider, for example, altering the climate. But nature has kindly provided us with natural experiments on the nearby planets. By examining them we can gain profound insights into how our planet works and how to make it work better.

Human beings have been as successful as they are because of our curiosity and intelligence, and also because of our spirit of exploration and adventure. At this point in human history, the surface of the planet Earth is fully explored. And at just this same moment, other worlds suddenly become accessible to us. I think it is in the deepest human tradition to seize that opportunity.

The Voyagers' Deep Space Exploration

By Ed Stone, as told to A.J.S. Rayl

In this selection, A.J.S. Rayl relates physics professor and researcher Ed Stone's story about working as project scientist for the *Voyager* mission. Stone says the *Voyager* mission, consisting of the *Voyager 1* and *2* spacecrafts, started in 1972 as a four-year program to explore Jupiter and Saturn, but it extended into a twelve-year mission to Uranus and Neptune as well. The scientist explains that the first seven years for the *Voyager* mission produced relatively routine discoveries, but in 1979, when *Voyager 2* reached Jupiter, the craft began sending tremendous amounts of data so quickly that scientists were unable to comprehend what they were seeing. In 1980 and 1981, when the craft reached Saturn, scientists had a similar experience, as well as five years later, when it reached Uranus.

Stone explains that the *Voyager*s made many more discoveries than expected because scientists had been unaware of how diverse the solar system was. He says that the unexpected discoveries were the most exciting, such as the fact that the 120 black spots on Jupiter's moon Io are volcanic calderas creating one hundred times more volcanic activity than on Earth, or the mysterious "kinks" in Saturn's rings that appear and disappear. According to Stone, the *Voyager* spacecrafts have completed their planned missions, but have enough electrical power to continue traveling for many years into the outer boundary of the solar system.

A.J.S. Rayl is an author and freelance investigative journalist whose work has appeared in *Air and Space*, the *Smithsonian, Astronomy, Scientist, Discover*, and *OMNI* magazines.

Ed Stone, as told to A.J.S. Rayl, "*Voyager* Is the Journey of a Lifetime," The Stories Behind the Mission: Personal Reminisces, www.planetary.org, 2002. Copyright © 2002 by The Planetary Society. All rights reserved. Reproduced by permission.

"Voyager* is the journey of a lifetime. It has been an incredible flight of discovery. I can't imagine a mission with more discoveries than *Voyager* had—it saw more new worlds for the first time than any mission and everything was different. And, it was happening in real-time.

My role as Project Scientist began in 1972 when Bud Schurmeier, who was the project manager, asked me to take the position. I was already part of a team that was hoping to get on the mission, so I was very interested in the science. We had a proposal that had been selected and today [in 1997] I'm the P.I. [principal investigator] for that cosmic ray instrument. The only question about the position was exactly what I would be doing, because being a project scientist meant a considerable, required load of administrative paper work. I had to ask whether or not it would be the best use of my time and interests.

While I was thrilled about the science, I was then (and am now) a professor at Caltech, who had my own research. I wanted to focus on the science and engineering issues. Bud understood that and worked with me to come up with a plan that would allow me to devote about 30% of my time in the role. Since no one had ever tried to do this before, this was an experiment. Up until then, project scientists had been fulltime and had done all of the administration, as well as the science.

Working Out an Arrangement

We worked out an arrangement that brought in a fulltime science manager who worked with me—James Long initially, and then Charles Stembridge and Pete de Vries—plus a group of experiment representatives. Eventually, there was also an assistant project scientist for Jupiter, Lonne Lane, and an assistant project scientist for the rest of the planets, Ellis Miner. The time I actually spent on the mission of course varied over the years—clearly during encounters it was 100%, but on the average it did command about 30% of my time over the years as planned. Ultimately, the key factors in my decision to accept the position were the mission itself, my desire to learn things, the support of both JPL [Jet Propulsion Laboratory] and Caltech, and Bud Schurmeier.

In the end, when I look back on it I say, 'Why did I decide to do it?'—well, Bud provided me with an opportunity to create a new kind of role. The creation of the position of a chief scientist who did the kinds of things I would do was not only a pioneering effort, it was a wonderful opportunity. You see, project scientists (or chief scientists) can have influence only to the extent that the project manager wants it and supports it, and I realized that Bud was going to be very supportive and that I was going to learn a lot from him and from the *Voyager* flight team. It was exactly the right decision. As I think back on it now, I cannot imagine it any other way. It has been a tremendous learning experience.

Big, Unexpected Surprises

There are so many discoveries that are highpoints and memorable moments, I'm not sure where to begin. I have 40 volumes (notebooks) of notes taken through all the meetings—of all the things we thought we knew, what we didn't know, what we thought we knew that wasn't that way.

The surprises were not the things that we expected to see—those are nice of course and important—but the most exciting things were the discoveries that were totally unexpected. Those were the biggest surprises. From a science point of view, when you see something that you had not expected at all—and which is not immediately obvious—like the volcanoes on Io or the fact that there are cracks on the ice crust on Europa, or the kinked F-ring [Saturn's outermost ring] at Saturn and the list goes on—that's when you have the opportunity to learn something new.

My first space mission was in 1961 and I had flown instruments on missions in 1965, '67, '69, '72, and '73—so I had lots of instruments in space, but they had all been on Earth-orbiters. The data would come in on a regular basis, always in a slow, steady stream, and if you made, say, one or two discoveries a year, you felt you had a wonderful program. Then—*Voyager.*

The Floodgates Opened

As we were cruising to Jupiter, we did have some of that same characteristic—a few discoveries per year on the way there for about a year and a half. So basically not much happened scientifically from 1972 until 1979—seven years. But with the encounter at Jupiter on March 5, 1979, the floodgates opened. I mean it just really was a flood. The data was coming in huge overwhelming deluges, faster than we could comprehend what we were seeing.

Each day, we would spend a few minutes looking at something and say—'Now can I understand this'—or not—and if not, we quickly moved on, because we knew the next day we were going to have even better data. It just got better everyday. It was a totally different kind of scientific experience than one normally has with the traditional sort of steady state, even science.

In 1980, another flood of data came from Saturn, and another one in 1981 at Saturn, and then five years later another flood at Uranus. There were so many more discoveries than any of us anticipated, because the solar system turned out to be much more diverse than any of us previously thought. To give you an idea— when this mission was first conceived back in the early 1970s, the best ideas of what the moons of the outer planets would look like were that they would be heavily cratered, ancient objects, much like our own Moon. When we saw them, they were each different. They each have a geologic life and very few are ancient cratered objects. The discoveries were just so much more than we could have imagined, because there's so much more diversity. The moons are diverse . . . every single ring system is different . . . the weather systems were not what we expected . . . the magnetic fields—a few planets we found have their magnetic poles down near the equator. Nobody even imagined that.

Volcanic Io

Just five days after the initial encounter at Jupiter we had the fly-by of Io and the images changed our thinking about this Galilean satellite. The Io story is interesting, because it shows how narrow our mindset was. About one month before Stan Peale, who with Pat Cassen and R.T. Reynolds, proposed that heating from tidal flexing [gravitational tug-of-war that occurs as Io moves closer to and farther from Jupiter] could melt the interior of a planet. He called to suggest that we might see a body that shows some evidence of this tidal heating effect. I sat down with Rudy Hanel, who was the P.I. for the infrared instrument, and said, 'Let's see if we can measure the increase of temperature of Io because of this extra heat.'

Well, while Peale and colleagues had calculated how much this tidal heating was going to provide, Rudy and I did our calculations for Io before the encounter and the numbers came out to be that the tidal heating [the heating of the interior of one planetary body

caused by stresses induced from the gravitational pull of another]
would provide additional heat of five microwatts per square centimeter. Now the question was 'How much heat does the Sunlight
provide?' It turns out that it provides 1000 microwatts per square
centimeter. So we said, 'There's no way we can measure that, it's
just too small.'

Then came the first images of Io. We looked at this thing and
we saw all these black splotches and people were saying, 'what is
this?' It doesn't look like anything we've seen before. Rudy came
in with his spectra a couple of days before our closest approach
and told the entire science group, some 100 people packed into
one room—that the temperature ought to be independent of the
wavelength at which we measure it in the infrared. But he said,
'We're not seeing that. We're seeing this very wave-length dependent temperature.'

Figuring Out Black Splotches

He offered three explanations. The first was that an unusual material or mineral on the surface had an absorption band that causes
the apparent change of temperature over this particular wavelength
range. He'd never seen anything like that before and it was hard
to imagine what that material might be, so that didn't seem like a
good explanation. The second suggestion was that there was a calibration problem with the instrument. But every other object he'd
measured had rendered the correct temperatures at all these wavelengths, so it was hard to view it as a calibration problem. The
third possibility was that there are different temperatures on the
surface of Io. That didn't seem too reasonable either, because it's
supposed to be quite cold. While we had known about the tidal
heating and we knew about the extra energy and we saw all these
black splotches—nobody jumped up and said, 'Hot spots!'

A few days later, the infrared instrument got close enough to Io
that they could look at one of these black lava lakes. It was room
temperature, while the surface itself was about 125 degrees above
absolute zero. The answer emerged—all the heat is not coming
out uniformly all over Io—it's coming out in these little black
spots. And, if the heat is coming out in little tiny areas, it has to be
much hotter. Ironically, this had been detected from Earth for
years, but it was just so hard to break out of the mindset that this
moon could not have volcanic activity. We just could not take the

leap. Then—the plume, which Linda Morabito (now Kelly) first saw. Bingo!

Suddenly, it all clicked together—we were seeing a world that had 100 times the amount of volcanic activity than Earth. We found ourselves saying, 'Okay, this is really different, and not just a little different—this is really different.' Today of course we know that all these black spots are volcanic calderas—all 120 of them.

Europa: Making Decisions and the Heart of Science

One of the key decisions that I had to make early on was in the se-lection of the trajectory of *Voyager 2*. *Pioneer 10* told us the radi-ation environment at Jupiter was 1000 times worse than we ex-pected, so we had to go back and redesign all the circuitry, and put shielding in and so on. That also meant we wanted to make sure that at least one of the spacecraft stayed far enough away from Jupiter to minimize risk. Remember, we had to get to Saturn—that was mission success. We wanted *Voyager 1* to go by Io, because we knew Io had a large interaction with Jupiter's magnetic field. Io is at 6 Jupiter radii out, where the radiation environments are intense.

The trade-off was to take some risk with *Voyager 1*, because *Voyager 2* which we would keep further out could always finish at Saturn if anything happened with *Voyager 1*. So we had to decide what to do with *Voyager 2* at Jupiter—do we go by Europa and look at it up close, or do we go by Ganymede again but go by be-hind it so we could look for a tenuous atmosphere by watching the sunset in the atmosphere, the solar occultation? There was some suggestion at the time from ground-based data that Ganymede might actually have an atmosphere.

It was a trade-off of one moon versus another, an atmosphere versus a surface. We would make discoveries either way. It was not only a question of what, but who. The imaging team would primarily make discoveries about Europa, and the ultraviolet team would primarily make discoveries about the atmosphere of Ganymede, so the outcome of the decision wouldn't involve the same people.

Such decisions—determining what gets discovered and who gets to discover it—are at the heart of science. This is right at the core of what scientists do. So this was a key decision. I decided that we should go for Europa to complete the quartet of the four Galilean satellites. I had not really appreciated at the time how really interesting Europa would be, because this decision-making was all done in the early 1970s, long before launch. We knew Europa had ice on its surface, but we didn't have any idea about tidal heating at that time.

The Galilean moons and Jupiter form a kind of miniature solar system and the idea was that we could view the fourth one and therefore accept the fact that we could not look for the atmosphere on Ganymede. It was clearly, in retrospect, the right decision. But that's an example of the decisions I had to make without knowing until after the fact that we'd done the right thing.

As I moved into this activity, I recognized that we needed to have a process so that everybody could accept whatever decision was made, even if it was not the one they would have made. I didn't feel that voting on things was the way. Somebody had to understand all the issues and that somebody was the chief scientist. I had to learn a lot in order to be able to make judgments ahead of time as to what things we would do—we couldn't do it all. Fortunately, there were plenty of discoveries for everybody involved and that made it a little bit easier for people to accept not getting to do everything.

Amazing Discoveries

Voyager started out as a four-year mission to Jupiter and Saturn, then turned into a 12-year mission to Uranus and Neptune, and with every new encounter there were startling surprises.

The kinked F-ring at Saturn was really a puzzle. *Pioneer 11* had discovered the narrow F-ring and we also knew from recent theories that there needed to be some sheparding satellites [pairs of moons in nearly the same orbit that keep ring material between them]. We found those and that was important—physics said they had to be there and they were. But I don't think anybody realized that the rings would be kinked—I mean, planetary orbits are ellipses—that's what everybody knows—and yet here is this kinked, multi-stranded ring. This was a huge puzzle. Now there are models that indicate how this can happen from the sheparding effects, but that was a big surprise.

Then, there are the spokes on Saturn's rings. We had always thought rings were made up of particles that orbit in a plane, but here are these features that are radial, that come and go. We still don't understand, by the way, what they are. That's a real opportunity for *Cassini*, which will observe them for at least four years.

At Uranus

At Uranus, the magnetic field was mind-boggling—the fact that it was so tilted. Miranda was another major surprise. This little world that is 500 kilometers across, one-tenth the size of the Galilean satellites, and yet has this very complex surface—how did that happen? It's a world that should have formed, rapidly cooled off and froze, but that's not what it did. Maybe it was broken up and reformed . . . maybe it never quite melted the last time it formed to reach a new equilibrium . . . there are answers still to be found from the very complex surface of this little world.

The last object we visited was Triton. We knew it was going to be interesting, because models had suggested it's a captured object, a twin of Pluto in many ways. But we had no idea how interesting. Being captured into a retrograde orbit it was going to have a lot of tidal flexing as it circularized its orbit. And it's like our Moon today—with one side always facing Neptune, so it no longer has all that tidal heating, but while circularizing its orbit it was being violently heated. Sure enough, that surface was different than anything else we had seen before. We saw icy volcanic calderas, which are basically not rock but ice that is as hard as rock when it's only 38 degrees above absolute zero—and polar caps that are made-up of frozen nitrogen, not water. Even now—even at 38 degrees above absolute zero—there are geysers erupting from the polar cap. So even that last object we saw surprised us.

The Story Is Still a Story

The fact that *Voyager* was a hit with the media and the public is important to note too. We had a story that built upon itself over a decade and longer and since this was a real-time mission we had the dedicated interest of the reporters. It took planetary exploration back into the mainstream.

Now [in 1997], 25 years after the launches, we still have enough electrical power for another 20 years or so. I'm sure as we head off to interstellar space—when we find the termination shock [boundary inside heliopause where supersonic solar wind particles are slowed to subsonic speeds by gas that exists between stars] and the heliopause, the outer boundary of the solar system, *Voyager* will reap even more surprises.

Looking back on this voyage now, if there is one thing we have learned along the way it is that nature is much more inventive than our imagination—and the journey of a lifetime is still not over."

The Hubble Telescope Has a Questionable Future

By Faye Flam

The Hubble Space Telescope, launched into Earth's orbit in 1990, realized the dreams of scientists who wanted a clearer view of space, free from Earth's atmospheric distortions. Indeed, the Hubble telescope has ten times more resolution than land-based telescopes, allowing it to view objects billions of light-years away. Scientists say that because of its views of galaxies never seen before and pictures of phenomena such as massive black holes, the Hubble telescope has changed scientists' ideas about the universe.

In this selection Faye Flam explains that the telescope's future is in jeopardy due to costly routine space shuttle servicing missions on the telescope. Although most scientists enjoy the benefits of the Hubble telescope, Flam says that many astronomers are ready to move on to the newer James Webb Space Telescope. The Webb telescope will be equipped with a mirror twice the size of Hubble's, allowing the telescope to see farther out in space. Because the new telescope will see in infrared wavelengths, rather than the visible light and ultraviolet wavelengths of Hubble, it will be capable of seeing the earliest stars formed in the almost fourteen-billion-year-old universe. Faye Flam is a journalist who writes for the *Philadelphia Inquirer* as well as for *Knight Ridder Tribune Business News.*

The Hubble Space Telescope which has opened new vistas on the universe, was supposed to keep working until 2010, when astronauts would go up in the space shuttle and bring it carefully back down to Earth. But plans have changed since the crash of the shuttle *Columbia* on Feb. 1 [2003]. Now, NASA is considering an earlier and more unseemly demise for the orbiting telescope, possibly forcing it to crash into the ocean before the end of the decade. And the next space telescope is already being designed, with capabilities to see well beyond Hubble's range.

The fate of the current telescope and the prospects for the new one may come into sharper focus this week [in July 2003]. A five-member committee charged with determining the Hubble's future is to hold a public meeting in Washington. . . . [As of early January 2004, the Hubble issue is still under review.] The big question is what to do with the Hubble.

Service Missions to Hubble Are in Jeopardy

The 13-year-old space telescope needs to be serviced every few years by astronauts who arrive aboard the space shuttle. But now, those repair visits are being reconsidered. NASA wants astronauts on space shuttle flights to be able to inspect and, if necessary, repair any damage to the shuttle's heat tiles and other components of its heat-protection system. The loss of heat shielding led to the *Columbia* disaster, which killed all seven astronauts aboard. Such inspection and repair is much easier if the shuttle is docking at the International Space Station. As it happens, the space station is part of the itinerary for all currently scheduled missions except the ones to Hubble. The shuttle can't carry enough fuel to visit both on the same mission. NASA could design special equipment for astronauts to check their heat protection on "stand-alone" missions, but the procedure would be more complicated, said Allard Beutel, a NASA spokesman.

Before the *Columbia* disaster, NASA had planned to send a shuttle up next year [2004] to make any necessary repairs and give the telescope two instruments—a new camera and a coronagraph, used to block bright light in order to see otherwise drowned-out objects, such as planets. Many astronomers had also been lobbying for one additional repair-and-upgrade mission, sometime around 2007, to extend the Hubble's useful life well beyond 2010. Now, though, the service visits are in jeopardy and the plan to recover Hubble at the end of its useful life is likely to be scrapped. The telescope will need to be visited at least one more time, because, left alone, it would eventually fall out of its orbit, possibly crashing into a populated area of Earth. One possibility is to conduct just one additional servicing mission in which astronauts would attach a propulsion system to guide the 43½-foot telescope into the sea.

Hubble's History and Accomplishments

The Hubble telescope grew from the long-held desire to get a clearer view of the heavens, above the distorting effect of the atmosphere. Several decades in the making, Hubble, named after U.S. astronomer Edwin Hubble, was finally launched in 1990, only to disappoint the world when the mirror turned out to have been manufactured incorrectly, blurring what should have been razor-sharp images. In a long-shot attempt to compensate for the problem, engineers created a set of "corrective optics," almost like eyeglasses. When astronauts installed the fix in 1993, the result was a stunning success—the telescope's vision became as clear as anyone had dreamed, and astronomy began to advance in leaps and bounds. (Hubble transmits its images and data via satellite to scientists in Baltimore, who view them on computer screens. The scientists use the same satellite system to send up instructions to direct the telescope's aim.)

"There are many things that changed our ideas about different parts of the universe," said Neta Bahcall, an astrophysics professor at Princeton University. Suddenly, astronomers could see into the centers of pinwheel galaxies, resolving individual stars. They realized that black holes with the masses of millions of suns were lurking there, whipping the stars around before swallowing them. Hubble was able to see further into space than other telescopes, and thus also further back in time. Objects billions of light-years away appear as they were billions of years ago, since it takes light that long for their images to get from there to here.

By training the telescope on a patch of seemingly empty space, astronomers found a dazzling array of galaxies never seen before, many of them so distant that they shine at us from the early universe, more than nine-tenths of the way back to the Big Bang. The universe looked different, more primitive, back then. "Galaxies were twisty, irregular things—they were smaller building blocks of today's galaxies," said Alan Dresser, an astronomer from the Observatories of the Carnegie Institution. Others trained Hubble on our planetary neighbors, mapping weather patterns on Neptune and storms on Mars. Most recently, Hubble made headlines for finding evidence of a 13-billion-year-old planet, the oldest known.

Hubble's Incredible Value

"Almost no matter what we look at—stars, galaxies, planets—it's been incredibly valuable to have Hubble look at it with 10 times better resolution than you could get from the ground," said Garth Illingworth, an astronomy professor at Lick Observatory in California. "Hubble's productivity is at its peak," said John Bahcall, an astrophysicist at the Institute for Advanced Study in Princeton and chair of a panel of experts assigned to evaluate Hubble's fate by October [2003]. Bahcall said that astronomers around the world are still competing for the ability to use Hubble. Only one in six proposals for Hubble projects is accepted. Bahcall said the panel would need to weigh the costly servicing missions with the potential to use the money for other projects, including Hubble's planned successor, called the James Webb Space Telescope, expected to be launched in 2011. "It costs a lot of money to maintain an army of people to potentially service the Hubble," he said. By abandoning Hubble early, "that money would find its way back into space science.". . .

The New James Webb Telescope

While astronomers almost unanimously love the Hubble, many are ready to move on and concentrate on the James Webb Space Telescope, named for the NASA director who oversaw the Apollo missions. The Webb telescope will have a mirror more than twice as wide as Hubble's, but the two telescopes see in different wavelengths—Hubble mostly in visible light and ultraviolet, Webb mostly in infrared.

The new telescope has been designed and the contract to build it was recently assigned to Northrop Grumman Space Technology, based in Redondo Beach, Calif. One of the main hopes for the Webb telescope is to see farther out in space and, consequently, further back in time. John Mather, who is chief scientist for the Webb telescope, says there's reason to hope that it will see back to the era when the very first stars lit up a previously dark universe. In theory, that happened several hundred million years after Big Bang—early in the 14-billion year history of the universe. These very first stars and galaxies are invisible to Hubble because their light has been "red shifted"—distorted by the expansion of space. Only a powerful infrared detector would have a chance to see them, and this is what the James Webb will carry.

Like Hubble, the Webb telescope should end up making discoveries of all kinds, said Mather, investigating black holes and planets that orbit faraway suns and the births of stars within galaxies. "We usually make progress faster when we get new equipment," Mather said. "People don't want to give up the golden goose with Hubble, but with James Webb we have the golden egg."

The Mars Observer Disappears

By John Travis, Christopher Anderson, and Jon Cohen

In August 1993 after a reasonably trouble-free eleven-month, 450-million-mile journey through space, the Mars *Observer* lost contact with scientists who were preparing the craft to enter the martian orbit. After desperately trying for days to unsuccessfully regain contact with the silent spacecraft, scientists numbly accepted that the first U.S. Mars mission in seventeen years was ended. The loss of the almost 1 billion dollar *Observer* was devastating to the scientific community. It impacted the lives and careers of hundreds of planetary scientists and researchers, some of whom had spent their entire professional careers working on the *Observer.* Scientists said that they had expected the *Observer* to provide more scientific data about Mars than all previous missions combined, enough to keep generations of scientists busy. The *Observer* was scheduled to spend a full martian year (687 days) probing and mapping the planet's surface, atmosphere, and interior, and would hopefully pave the way for manned missions to Mars. Because of the importance of the mission, some scientists were immediately ready to rebuild the Mars spacecraft.

John Travis is a science journalist specializing in biology and biomedical research for *Science News*, a weekly science magazine. He formerly worked as a New England correspondent for *Science* magazine, an international weekly journal of the American Association for the Advancement of Science. Christopher Anderson is a writer for *Science* magazine. Jon Cohen is a contributing editor for *Science* magazine. Cohen has specialized in biomedicine and vaccines for fifteen years and is one of the world's leading AIDS reporters. He has written for numerous publications, including the *Atlantic Monthly*, the *New Yorker*, the *New York Times*, the *Washington Post*, the *Smithsonian*, and the *New Republic.*

John Travis, Christopher Anderson, and Jon Cohen, "Mars *Observer*'s Costly Solitude," *Science*, vol. 261, September 3, 1993, p. 1,264. Copyright © 1993 by the American Association for the Advancement of Science. Reproduced by permission.

As Mars *Observer* closed in on the red planet [in August 1993], Alden Albee, the mission's project scientist, thought about calling *Science* to discuss the possibility of publishing a report of the spacecraft's observations when it had circled the planet for 60 days. Something made him hesitate. "I was just being cautious and decided to wait until it was in orbit," the dean of graduate studies at Caltech recalls. To Albee's horror, his restraint proved more than justified.

On Saturday evening, 21 August [1993], ground controllers at the Jet Propulsion Laboratory (JPL), which manages the mission, lost contact with the spacecraft, and, despite their frantic efforts, it has not been heard from since. By last weekend, JPL had nearly run out of ideas to shock *Observer* into communicating with Earth, and controllers contemplating a "listen-only" vigil in the slim hope that the spacecraft would wake up and call home.

A Disastrous Impact

The abrupt end of the first U.S. Mars mission in 17 years, coupled with a string of other recent problems in the space program, could not have come at a worse time politically for the National Aeronautics and Space Administration (NASA). The space agency's budget for fiscal year 1994 will be decided on Capitol Hill in the next few weeks, and NASA needed a major scientific and public relations triumph to help persuade Congress to back a big increase in funds. The impact on the hundreds of planetary scientists whose lives and scientific careers have been intimately intertwined with the spacecraft's progress is more immediate. "It's difficult to work on something for 10 years, and expect to work on it for another 5 years, and have it disappear," says Philip Christensen of Arizona State University, one of *Observer*'s principal investigators. "[In] the next 3 to 6 months, most if not all the people working with me will be looking for jobs," says University of Arizona's William Boynton, team leader for *Observer*'s gamma ray spectrometer.

Scientifically, the loss is no less disastrous. Researchers were hoping *Observer* would help answer mysteries about Mars' geologic history and its turbulent atmospheric circulation. And *Observer*'s detailed mapping of Mars' surface was supposed to pave the way for future NASA and international forays to the planet—perhaps even a manned mission someday. "This is the kind of data that would allow us to really understand Mars," says Albee. Indeed, he and his colleagues in the planetary science community believe the mission was scientifically so important that they are urging NASA to find some way to repeat it.

The Scientists Involved with Observer

With a total cost approaching $1 billion, *Observer* may seem like the epitome of "Big Science," but the mission actually combined many smaller observational projects, and its apparent demise has touched the careers of hundreds of individual researchers. Albee, who says *Observer*'s deathly silence has left him "numb," is one of more than 100 investigators who have been planning the mission for the past decade. Beyond that core group, says Albee, are perhaps 500 scientists who would have worked with the data. And behind those front ranks is an army of postdocs, graduate students, and future investigators whose careers might have been shaped by the wealth of information from *Observer*. "I run a team of 25, with seven graduate students and five undergrads. We've been trying to teach the next generation," says Christensen.

What makes the tragedy of Mars *Observer* even more cruel is how close the mission came to succeeding. The team behind the spacecraft's radio science experiments had gathered at JPL on Monday and Tuesday, as the spacecraft was scheduled to go into orbit, to make final preparations for the onslaught of data they had expected, but *Observer*'s silence turned the gathering into a wake. "It's difficult to concentrate. It's depressing to even talk at the moment. These kinds of missions demand a long-term effort from a lot of people," Stanford University's David Hinson told *Science* after the meeting.

Though it was small comfort at the time, Hinson said he may be one of the more fortunate members of the *Observer* team, because he has other projects to fall back on. NASA geologist James Garwin, who worked on the spacecraft's laser altimeter experiment, says: "Basically, it's my entire scientific professional career."

Then there's the plight of Michael Malin, principal investigator for the spacecraft's camera system. Originally, NASA had no plans to include any cameras on *Observer*, believing the *Viking* missions in the mid-1970s had provided more than enough pictures. But Malin fought fiercely to change the agency's mind. He developed *Observer*'s innovative camera system and even started his own company to handle the operation of the instrument and the analysis of the images. Now his company may be out of business. The experience has been emotionally wrenching: "A TV reporter will go up to someone who's just lost children in a fire and say, 'How do you feel?' Well, that's how I feel. Like TV reporters are shoving mikes in my face. I don't get sad. I get angry."

Lost Scientific Data

Beyond the personal anguish, the apparent loss of Mars *Observer* is devastating to the planetary science community. It would have provided more data about Mars than all previous missions combined. After moving into a low-altitude, near-polar orbit, the spacecraft was scheduled to spend a full Martian year (687 days) probing and mapping the planet's surface, atmosphere, and interior. *Observer* would have delivered eagerly awaited answers about Mars' geology, geophysics, weather, and climate. Among the key questions: Has Mars ever had an Earthlike atmosphere? Where is the planet's water hidden? Are there active volcanoes on its surface? How do planet-wide dust storms develop? Does Mars have a molten core and consequently a magnetic field?

To address such issues, *Observer* was loaded with seven scientific experiments:

- Gamma Ray Spectrometer (GRS)—By measuring the intensities of gamma rays from the Martian surface, GRS would have provided a detailed planet-wide analysis of the elements on the planet's surface.
- Mars Observer Camera (MOC)—Two low-resolution cameras would have provided daily weather and surface maps and a high-resolution camera could have targeted sites of special interest.
- Thermal Emission Spectrometer (TES)—like GRS, TES would have analyzed the mineral content of the Martian surface. Other goals included observing movement of polar ice caps and the distribution of atmospheric dust.
- Pressure Modulator Infrared Radiometer (PMIRR)—By detecting infrared radiation from the atmosphere itself, PMIRR would have provided measurements of atmospheric pressure, temperature, water vapor, and dust that could have led to global atmospheric models of Mars.
- Mars Observer Laser Altimeter (MOLA)—MOLA was designed to develop an extremely detailed topographic map of

Mars by bouncing short pulses of laser light off the surface.
- Radio Science (RS)—By measuring minute Doppler shifts in *Observer*'s radio signals back to Earth, investigators would have built up a gravity map of Mars as well as explored the planet's atmosphere.
- Magnetometer and Electron Reflectometer (MAG/ER)—This experiment might have settled whether Mars has a magnetic field and, if not, whether it had one in the past.

More important, perhaps, than the individual data from each of these instruments would have been the opportunity to combine all these results over a full year, providing a cross-checked, dynamic portrait of the red planet. *Observer* "was not designed for just one little question. This represented the possibility of accruing an incredible dataset that would have been used for generations of scientists. . . . We were going to end up with better global data for Mars than we have for Earth. Atmospheric profiles, cloud patterns—all these things would have systematically been done and they would have been in a database accessible to the world community," says Albee.

Looking at the Future

In addition to raining a torrent of data of immediate interest down on Earth-bound researchers, Mars *Observer* was expected to provide a key stepping stone toward future trips to the planet. Russian scientists were hoping to use *Observer*'s maps to help choose sites to land instruments on the planet's surface in missions scheduled for 1994 and 1996. And they were planning to use a French-built relay system on *Observer* to transmit back up to 10 times as much data as their own orbiter could handle.

Observer's data would also have fed directly into NASA's Mars Environmental Survey (MESUR), a proposed mission to land as many as two dozen geophysical probes on different regions of the planet. Although photographs from the *Viking* mission can pinpoint relatively safe sites to land, *Observer*'s detailed mapping would have allowed NASA to pick from among many more targets and would have indicated which ones would be of greatest scientific interest. "Without the measurements of geochemistry [from *Observer*], intelligently choosing the sites will be difficult," says Boynton.

Considering the importance of Mars *Observer*, Boynton and his colleagues in the planetary science community are already urging that the mission be repeated. The next launch windows for a return to Mars are late 1994 and 1996, which means that decisions would have to be made quickly. Rebuilding the spacecraft itself might not pose a great problem: Once NASA realized Observer would be a one-of-a-kind mission, it bought backups for the major components. There are also spare parts for most, if not all, the instruments. For example, says Christensen, "we can rebuild our [thermal emission spectrometer] in 6 months to a year, given the funding."

Funding Is Biggest Problem

Funding, of course, is the biggest problem. A repeat *Observer* mission should cost significantly less—perhaps 20% to 30% of the original mission's cost says Boynton—but in the current funding climate, NASA will be hard pressed to muster political support for funding the space station and also get additional money to try *Observer* again. Nevertheless, *Science* has learned that NASA has inquired at Lockheed, which is now managing the Russian Proton launch system, about the availability and feasibility of a rocket to launch a carbon copy of *Observer*.

Another alternative under consideration involves the Clementine series of military satellites that were scheduled to test technology for the Ballistic Missile Defense Organization. NASA declines to comment publicly, but space agency officials have asked whether a small fleet of Clementines, outfitted with one or two of *Observer*'s instruments, might be ready by November 1994.

For the moment, however, the shaky hope of another foray to Mars offers little solace to a scientific community that is reeling from the loss of *Observer*. "With planetary science, there's a 10- to 15-year gap between experiments. That's what really hurts. Most experimental scientists can redo an experiment in a few months," says Christensen. Still, he and his colleagues aren't about to give up. "Planetary scientists, because it's a risky business, are an incredibly optimistic group. The process of looking to the future is already starting."

The Sounds of Silence: Chronology of Despair

Ten days after they lost contact with Mars *Observer*, bleary-eyed ground controllers at the Jet Propulsion Laboratory (JPL) were left wondering . . . whether they would ever know what happened to their errant spacecraft. As *Science* went to press, they had run out of strategies for trying to find out, but they were still clinging to a faint hope that the spacecraft might yet rise from the presumed dead.

The nightmare sequence of events began on Saturday evening (9:00 P.M. EDT), 21 August, as the spacecraft readied itself to enter the planet's orbit after a relatively trouble-free, 11-month, 450-million-mile journey through space. *Observer* would face its most critical maneuver 3 days later when the spacecraft had to fire two of its four large thrusters to decrease velocity and allow Mars' gravity to pull it into a planetary orbit.

In preparation, *Observer* needed to pressurize its fuel tanks. To do this, valves are opened with a small explosive device; the device also delivers a short, small shock to the whole spacecraft. Before the explosion, which was scheduled for Saturday night, *Observer* followed instructions previously loaded into an onboard computer and shut down its transmitters to protect them from the shock. After pressurization had presumably occurred, JPL controllers sent commands telling *Observer* to begin transmitting again. They were greeted with a surprising silence.

No Sense of Disaster at First

For the first few hours, there was no sense of disaster at JPL. *Observer* had experienced minor communications problems several times before and so had many other planetary missions. Indeed, the *Magellan* mission to Venus suffered a blackout of more than a dozen hours in its early days around Venus and many had feared the spacecraft was lost. Moreover, JPL controllers were buoyed by the fact that in their last message to *Observer* on Friday they had, as a planned precaution, included a complete set of instructions for entering orbit. Mission controllers worked around the clock, sending commands directing the spacecraft to turn on its transmitter, to switch from its high-gain antenna to one of the three low-gain backups, and to change to a backup computer. As the hours passed with no success, concern grew.

On Monday, a day before the scheduled orbit insertion, tests on a simulator at JPL known as the Verification and Test Lab suggested an onboard clock, which orchestrates the sequence of operations aboard the spacecraft and is necessary for any computer operations, might be stuck. JPL sent commands to activate a backup timing mechanism. Silence prevailed.

Observer team members were still putting on a brave face on Tuesday. "We think the spacecraft is operating and just is not talking with us," one mission manager told *Science*. By Tuesday evening, however, it was getting harder to remain optimistic. *Observer* was programmed to call back after thrusting itself into orbit that afternoon, but, with live NASA television documenting the events, no signal came through. That left ground control with no idea whether *Observer* had been somehow destroyed, was circling Mars in silence, or had failed to fire its thrusters and sped past the planet entirely.

Other Scientists Spring into Action

While JPL was struggling to regain contact with its missing craft, other scientists sprang into action. Researchers at Cornell frantically calculated whether they could bounce radar off the craft to see where it was. Their disheartening conclusion: Only if *Observer* had sped by Mars might they be able to detect the spacecraft a year or so from now when it had looped around the sun and was again near Earth.

NASA scientists interrupted their search for extraterrestrial intelligence to look for signals from *Observer.* Michael Klein, who directs NASA's Sky Survey, says his group took raw data from two NASA antennas tracking where *Observer* should be and, with their specialized computers, scanned more than 2 million different radio frequencies for any weak signals. The Sky Survey's own 34-meter antenna was also pressed into service to conduct a small search pattern around Mars in case the spacecraft had been somehow knocked from its expected position, perhaps by a thruster misfiring. "It may be a longshot, but it sure would be exciting if we found it," Klein told *Science*.

Perhaps the best chance of knowing whether *Observer* had at least gone into orbit fell prey to poor weather. Astronomers at NASA's Infrared Telescope Facility on Hawaii trained their instrument toward Mars, thinking they might be able to see the heat flare from *Observer* if it had fired its thrusters. As the crucial 30-minute burn approached, however, the island's normally sunny skies clouded over. The telescope couldn't even make out Mars, let alone see the faint signature of rocket motors.

On Wednesday, controllers at JPL waited and held their breath as another critical milestone approached. At 5:56:53 P.M. EDT, it had been 5 days since JPL could confirm *Observer* had received their messages. The spacecraft contains a "command loss timer," which is reset every time instructions are received from Earth. But if *Observer* hears nothing for 5 days, the timer prompts it to look for Earth and point its low-gain antennas homeward. The milestone passed with no contact. As the weekend approached, JPL was forced to try more drastic, last-ditch efforts, even sending commands to "cold reboot" the onboard computers. Nothing worked. Their options virtually exhausted, JPL could only listen and hope the craft could overcome its problems.

Transistor Failure May Have Been Problem

As for what might have gone wrong, a leading theory has become transistor failure. JPL acknowledged . . . that both the master and backup clocks on the spacecraft depended on a suspect lot of transistors. In June [1993], a clock failure had delayed the launch of a weather satellite and an investigation later found that the device's transistors, the same as used on *Observer*, contained a weak weld that was susceptible to breaking. While there are no data to confirm the notion, the jolt of pressurization may have prompted the failure of both clocks, dooming the craft.

NASA has named Timothy Coffey, director of research at the Naval Research Laboratory, to head an investigation into the loss of communications with *Observer.*

How Observer Became a Billion-Dollar Mission

Mars *Observer* provides an object lesson in how budget constraints, and scheduling delays, can drive a mission that started out with a modest price tag into the billion-dollar range. The final tally for the ill-fated spacecraft—including development, construction, launch, and operation for 11 months—now stands at $845 million.

That hefty price tag certainly was not what NASA envisioned in the early 1980s when the agency's solar system exploration committee laid out plans for revisiting Mars. The panel recommended a quick series of modest missions, part of a new *Observer* class of spacecraft, that would save money by carrying just a few routine instruments and using designs of current Earth-orbiting satellites instead of being designed from scratch. The first such craft might cost around $200 million, but the price of subsequent *Observers* would drop to $150 million, officials hoped. In many ways, the approach foreshadowed the "smaller, cheaper, faster" *Discovery* missions NASA administrator Daniel Goldin now supports, says David Morrison, who chaired the NASA committee.

Congress Rejected Plans

The series, however, never got off the ground in that form. Rejecting NASA's proposal, Congress and the Office of Management and Budget approved planning for only a single mission, which eventually became Mars *Observer.* With only one shot at the red planet, planetary scientists started to pile instruments originally planned for multiple missions on a single platform, at the same time making them increasingly sophisticated. Late in *Observer's* conception, for example, NASA added a camera system that eventually cost more than $20 million. Moreover, since there would not be a string of *Observer* spacecraft, the cost of Mars *Observer* was driven up $60 million by the need to include backups for everything in case of a launch disaster.

These escalating hardware costs, according to NASA officials and outside analysts, were accompanied by even bigger cost increases caused by the agency's dependence on the space shuttle as launch vehicle. *Observer*, for instance, had to be built for a man-rated vehicle, a safety demand that significantly increases costs. In addition, a special booster was needed to send the spacecraft from shuttle orbit to Mars.

Then came the explosion of *Challenger* in 1986. *Observer* had not been scheduled to launch until 1990, but the shuttle disaster rocked NASA's budget plans and meant some drastic consequences for the Mars mission. Among them were a 2-year delay in launching, a decision to send the *Observer* up on a Titan III rocket, and a resulting design reconfiguration. Those events, says NASA, added $93 million to Mars *Observer's* final cost. "Until the *Challenger* accident came along, we were in reasonably good shape. That was a very big cost hit," says Geoffrey Briggs, NASA's director of solar system exploration from 1983 to 1990.

In the final analysis, where did the $845 million go? According to NASA headquarters, the spacecraft and instruments, including spares, and the cost of integrating them add up to $511 million.

Costs attributed to the 1992 launch, including the price of the Titan III itself, total $282 million. To top it off, throw in another $30 million in operational costs over the last 11 months, approximately $15 million for expenses like tracking the spacecraft with NASA's Deep Space Network of antennas, and $7 million for NASA's contribution to the Mars Balloon Relay, and there's not much change from $1 billion.

Rovers Reveal That Mars Was Once Wet

By Marsha Walton

In January 2004 two space rovers, *Spirit* and *Opportunity*, success-fully landed on Mars. According to Marsha Walton, in the following selection, scientists were overjoyed with the information that the two rovers transmitted to NASA's Jet Propulsion Laboratory in Pasadena, California. Using instruments on the rovers to study martian rocks and soil, scientists were able to verify that Mars had had water on it at one time. However, scientists could not determine when water had existed on Mars until further examination of rock and soil samples. Walton said that the planet's watery history insinuated a more excit-ing possibility of life on Mars. Marsha Walton is a news correspon-dent and producer for CNN News, a worldwide news and informa-tion source that has global bureaus and a team of four thousand news professionals.

"**W**e have concluded the rocks here were once soaked in liquid water," said Steve Squyres of Cornell Univer-sity. He's the principal investigator for the science in-struments on *Opportunity* and its twin rover, *Spirit.*

"The second question we've tried to answer: Were these rocks altered by liquid water? We believe definitively, yes," Squyres said.

Squyres and other NASA officials made the announcement at NASA headquarters in Washington, after several days of giving tantalizing hints that something significant had been discovered.

"Three and a half years ago, in July 2000, we were on stage here to talk about sending two rovers to get evidence of past wa-ter. NASA and its international partners have turned those dreams to reality," said Ed Weiler, NASA associate administrator for space science.

Scientists used instruments on board the golf cart–sized rovers to study the composition of the rocks and soil on the planet. The rocks' physical appearance, plus the detection of sulfates, make the case for a watery history, and more important, an environment that could have been hospitable to life.

While reporters pushed the scientists to come up with a "when" for the existence of water on Mars, Squyres said it was very difficult to infer an age simply by looking at pictures. He said a physical examination of samples would be the only way to get close to a time frame.

Squyres did offer a couple of scenarios on what might have happened that led to the current discoveries:

One is that there was a volcanic eruption, possibly many eruptions, and volcanic ash settled out onto the Martian surface. Subsequently, water could have percolated through the ground, altering the ash to the chemical composition it has today.

Another possibility, said Squyres, is that there was a salty sea at the Meridiani Planum location [where rovers traversed], perhaps with currents, possibly even waves. As the water evaporated, the salt would settle out.

"Both are fundamentally possible," said Squyres. "But we may never know."

Two Rovers Land on Mars

Spirit and *Opportunity* were sent to opposite sides of the planet with the possibility of investigating different types of terrain. *Spirit*, the first rover to arrive on January 3 [2004], landed near the Gusev Crater, which may once have held a lake.

But geologists and other researchers at NASA's Jet Propulsion Laboratory in Pasadena, California, were thrilled when they saw the possibilities surrounding *Opportunity*, which landed three weeks later. It landed inside a small crater in the Meridiani Planum, one of the flattest places on the planet. And its landing site was within driving distance for the spacecraft to reach an exposed slice of bedrock.

Since its landing January 25 [2004], *Opportunity* has used the same tools as a human field geologist would to determine the chemical contents of the rocks. Using an alpha particle X-ray spectrometer, a device that can identify chemical elements, scientists have identified a high concentration of sulfur in the bedrock.

Another instrument on board, a Moessbauer spectrometer, has detected an iron sulfate mineral known as jarosite. From their knowledge of rocks on Earth, scientists say rocks with as much salt as this Mars rock either formed in water, or had a long exposure to water after they were formed. The scientists say these rocks could have formed in an acidic lake or even a hot springs.

Scientists say the case for a watery past is further strengthened by the pictures taken by the rovers' panoramic cameras and its microscopic imager. One target rock, named "El Capitan," is filled with random pockmarks. Geologists say a texture like that comes from sites where salt crystals have formed in rocks that have sat in salt water.

Scientists say they have gained other clues from the physical appearance of the rocks. They see a pattern called "crossbedding," which is often the result of wind or water moving across the rock's surface.

So what is ahead for the final few weeks of the rovers' operations on Mars?

"We need to take a close look at the outcropping, and broaden our view to get a better understanding of the geology of the region, which is about the size of Oklahoma," said Joy Crisp, project scientist at the Jet Propulsion Lab. She said there are also plans to drive about 740 meters east to a crater that has been nicknamed "Endurance."

And in the longer term?

The Future for Mars Missions

"It's clear we have to do a sample return, both for the scientific side and in preparation for human landing," said Weiler. He said future Mars missions would also include miniaturizing equipment, and landing equipment that would help prepare for the eventual landings of humans. That might include tests for toxicity in the soil, and to determine if there are any materials that humans might find useful when they do arrive.

The cost of the two rover missions is about $820 million. With solar panels and lithium-ion battery systems aboard, each rover is expected to function and communicate with Earth for about 90 Mars days, known as "sols." That's equivalent to 92 Earth days.

CHAPTER 4

Current Controversies and Outlooks in Space Exploration

Human Space Explorers Are Superior to Robots

By Robert Zubrin

In this selection, aerospace engineer Robert Zubrin supports an argument for continuing to send humans to explore space. Zubrin believes that robots cannot match human intuition, versatility, ingenuity, and common sense in the field of space exploration. He says that robots are not capable of performing tasks such as paleontological digs that will be necessary on future planetary missions.

Because Zubrin believes that NASA has become less significant over the last thirty years due to a lack of purpose, he says that NASA should focus on the exploration of Mars as well as on sending humans to the red planet. Sending humans to Mars, Zubrin believes, would inspire young people to become scientists and engineers in the field of space exploration. In addition, it would prove human's capability in space and stimulate more daring ventures into space.

Robert Zubrin is an aerospace engineer and inventor of several concepts for space propulsion and exploration. Zubrin is president of Pioneer Astronautics, a research and development company, as well as founder and president of the Mars Society. He is an author of over one hundred published technical and nontechnical aerospace papers and author of the books *The Case for Mars, Entering Space, First Landing*, and *Mars on Earth.*

President [George W.] Bush has called for the human exploration of space. His vision changes the orientation of the American manned spaceflight program from one of observing and gathering data on the human experience in space—the medical effects of zero-g and so forth—to a program of going into space to travel across it, to explore worlds.

The president's plan is a step in the right direction, because it gives NASA a much-needed goal. The main reason why NASA's level of achievement in the last three decades has paled in comparison to NASA's level of achievement from 1961 to 1973 is that, in that earlier period, President Kennedy set a clear goal: Reach the moon within a decade. With that specific mission, NASA did *Mercury, Gemini, Apollo*, and *Skylab*. It did a host of robotic missions. It developed virtually all the space technologies that we have today, and all the major American space institutions.

But without a goal, NASA's level of achievement has declined, even though NASA's average budget in the 1990s was similar (in inflation-adjusted dollars) to the average NASA budget from 1961 to 1973. The problem has not been a lack of money but an absence of purpose.

The Right Goal for NASA

So the question arises: What is the right goal for NASA? The right goal is Mars for the following three reasons: Mars is where the science is, Mars is where the challenge is, and Mars is where the future is.

First, the science. A multitude of issues in different scientific disciplines—planetary geology, meteorology, seismology, and other fields—could be answered by putting humans on Mars. But the central questions involve life: Was there, or is there, life on Mars? And if so, what is the nature of that life?

To uncover whether life evolved from chemistry on Mars is the critical experiment for knowing whether the evolution and development of life from chemistry is a general phenomenon in the universe, wherever appropriate physical and chemical conditions exist. Mars appears to have had liquid water for a significant period of time—longer than it took for life to appear in the fossil record on Earth after there was liquid water here. So it's the Rosetta Stone for letting us know the answer to the question of life.

It's also the key for discovering whether all life has to have the same form as life on Earth, since all life on Earth is the same at the biochemical level. It all uses the same RNA and DNA methods of replicating information and so forth—but perhaps it does not have to. Going to Mars can help us find out.

Advantages of the Human Explorer

On Earth, fossil-hunting involves hiking long distances through unimproved terrain, and climbing up steep hillsides or cliffs. It involves digging and pickax work, as well as delicate handiwork, like carefully splitting open shales edgewise to reveal the fossils that have been trapped between the pages of rocks pasted together. This is far beyond the ability of robotic rovers like *Spirit* and *Opportunity*. If you took one of these robots to a paleontological dig on Earth, the researchers might use it as a platform for putting coffee cups on.

It is true that wearing a spacesuit greatly reduces your situational awareness. Obviously you don't have a sense of smell or a direct sense of touch. But you do retain the ability to pick up samples and manipulate them, and the ability to break rocks open and look inside them. You are able to walk back and forth, looking down at rocks and taking in with your eyes the equivalent of millions of high resolution images.

The human explorer can follow up on all sorts of intuitive clues and observations. Out of the thousands of rocks he has glanced at and the hundreds he has looked at more closely—perhaps he brings ten samples back into the habitat. There he can look at them with a hand-lens; he can thin-section them; he can examine them under a microscope.

Robots Have Limitations

Two hundred years after Lewis and Clark, there is not a robot on this planet that you can send to the grocery store to pick up a bag of unbruised apples. If they can't do a trip to the grocery store, how can they explore a planet? How can robots match the intuition, versatility, ingenuity, and common sense of the human explorer?

Now, I'm not putting down robots. It is excellent and important to do robotic missions. But a robot explorer on the surface of a distant planet simply cannot duplicate what a human explorer could do. To find out whether there is life on Mars, we're going to have to set up drilling rigs, drill down into the ground, sample the water, bring it into the lab, and examine the samples under a battery of tests, with a scientist who can react flexibly to the data, in consultation with other scientists back on Earth.

Remember, we've sent robots to Mars to search for life before. In 1976, we sent *Viking* and did four tests on the Martian soil to determine if there was life. Three suggested there might be life; the other was negative. The meaning of these experiments is still being debated. *Viking* asked Mars, "Do you have life?" Mars said, "Maybe. Please rephrase the question." If there were humans there, they could immediately have rephrased the question by performing additional experiments. This is the superiority of human exploration.

Mars Is a Challenge

The second reason to go to Mars is the challenge. It is the chance to do something heroic, to advance humanity on the frontier. A humans-to-Mars program would be a challenge to our entire society. In particular, it would be an inspiration to the next generation: "Learn your science and you can pioneer a new world. Develop your mind and you can be a hero for humanity—doing something that has never been done before, seeing things that no one has seen before, building where no one has built before."

This is the challenge that the youth of my generation got from the *Apollo* program. As a result of that challenge, I became an engineer. As a result of that challenge, the number of scientists and engineering graduates in this country doubled at every level: high school, college, Ph.D.

And what did those people end up doing? Some went into aerospace, but most of them went off into other scientific ventures: they engaged in medical research, they built Silicon Valley, they created the economic boom of the 1990s. Those 40-year-old techno-nerd billionaires of the 1990s were the 12-year-old boy scientists of the 1960s. It's an investment: If we go to Mars, we'll someday get the payoff that comes from challenging people in a serious way, and by being a society that values great scientific and human achievements.

The final reason to go to Mars is the issue of the future. Imagine you lived 50,000 years ago in Kenya, along with the rest of the human race, and received a proposal from someone who thought humans should colonize Europe or Asia. The skeptics would have

said: "Those places are impossible to live in. It's much too cold."
If they had robotic probes, they might have sent them to confirm
these assumptions with more precision: "Our robotic probes show
you could not survive a single winter night in Europe."

But people were able to colonize Europe with the aid of technology: clothing, houses, fire. It is on the basis of our technological ingenuity that humans have left our natural habitat, the
Kenyan Rift Valley, and transformed ourselves into a global
species with hundreds of nations, languages, and cultural tradi-
tions. There has been a vastly richer human experience as a result
of the human willingness to leave the known in order to explore
and master the unknown. And this is the challenge that Mars holds
for us today.

Mars is not just an object of scientific inquiry. It is a *world*. It
is a planet with a surface area equal to all the continents of the
Earth put together, with all the resources needed to support not
only life but technological civilization, should we choose to exer-
cise our creativity sufficiently to make that possible.

A New Branch of Civilization

If we do what we can in our day, which is establish that first human foothold on Mars, then five hundred years from now there
will be a new branch of human civilization living there. Perhaps
many new branches of human civilization will flourish on Mars,
with their own cultures, their own languages yet unspoken, their
own novel ideas on human social organization, their own tradi-
tions of heroic deeds, and their own manifest contributions to technology and invention.

And that is something wonderful. That is something enor-
mously valuable. I wonder if we can even put a price tag on help-
ing to give birth to a new branch of human civilization, one that
contributes in unimaginable ways to human progress and the human story.

And not only that, but a branch of civilization whose development shows us that we have the capability to do such things, the
capability to engage in yet greater ventures, more daring ventures,
further out, toward an unlimited future. And that is the reason why
humans should go into space.

Human Accomplishments on the International Space Station Will Be Inconsequential

By Robert L. Park

In the following congressional testimony, physics professor Robert L. Park argues that the experiments conducted on the International Space Station (ISS) will be unimportant. Park says that the ISS experiments will be mere extensions of experiments that have been done on the space shuttles over the past twenty years. Although the ISS is an orbiting laboratory for the study of a microgravity (weightlessness) environment, Park says scientists have already shown that this environment is deleterious to human health. Furthermore, scientists are scheduled to perform protein crystal experiments on the ISS, which Park says scientists have already done as well.

Park argues that robots should be used instead of humans in space exploration. He points out that robots have already visited every planet except Pluto and believes that by the time humans land on Mars, robots will have long finished exploring the planet. Robert L. Park is a professor of physics at the University of Maryland as well as director of public information in the Washington office of the American Physical Society. Park is the author of the book, *Voodoo Science*, and *What's New*, an electronic newsletter.

Robert L. Park, testimony before the U.S. Senate Commerce, Science, and Transportation Committee, Washington, DC, October 29, 2003.

A space station once seemed to be an inevitable step in the conquest of space. From such a platform it would be possible to relay communications around the globe, track weather systems, detect military movements, provide navigational assistance to ships and planes, and study the heavens free of atmospheric distortion. All these things and more are now done routinely using unmanned satellites, and these robotic spacecraft are doing the job far better and far more cheaply than would ever be possible with a manned space station.

The International Space Station is an orbiting laboratory for the study of a microgravity environment. There are two quite separate justifications for a microgravity laboratory: One is to examine the biomedical effects of extended human exposure to microgravity; the other is to determine whether microgravity offers any advantage in manufacturing.

There had been speculation that certain manufacturing processes that are difficult or impossible on Earth might be easier in microgravity. For most manufacturing processes, however, gravity is simply not an important variable. Gravitational forces are generally far too weak compared to interatomic forces to have much effect.

The Myth Surrounding Protein Crystals

A possible exception was thought to be the growth of molecular crystals, specifically protein crystals. The structure of protein molecules is of enormous importance in modern medical research. Protein crystals make it possible to employ standard X-ray crystallographic techniques to unravel the structure of the protein molecule. It had been speculated that better protein crystals might be grown in zero gravity.

Unlike the interatomic forces within a molecule, molecules are bound to each other by relatively weak forces; the sort of forces that hold water droplets on your windshield. Gravity, it was supposed, might therefore be important in the growth of protein crystals. Indeed, in the days following the *Columbia* tragedy, NASA repeatedly cited protein crystal growth as an example of important microgravity research being conducted on the shuttle. NASA knew better. It was 20 years ago that a protein crystal was first grown on Space Lab 1. NASA boasted that the lysozyme crystal was 1,000 times as large as one grown in the same apparatus on

Earth. However, the apparatus was not designed to operate in Earth gravity. The space-grown crystal was, in fact, no larger than lysozyme crystals grown by standard techniques on Earth.

But the myth was born. In 1992, a team of Americans that had done protein crystal studies on Mir [the Russian space station] commented in *Nature* (26 Nov 92) that microgravity had led to no significant breakthrough in protein crystal growth. Every protein that crystallizes in space also crystallizes right here on Earth. Nevertheless, in 1997, Larry DeLucas, a University of Alabama at Birmingham chemist and a former astronaut, testified before the Space Subcommittee of the House that a protein structure, determined from a crystal grown on the shuttle, was essential to development of a new flu medication that was in clinical trials. It simply was not true. Two years later *Science* magazine (25 June 99) revealed that the crystal had been grown not in space but in Australia.

Biologists Call for Program's Cancellation

Meanwhile, the American Society for Cell Biology [ASCB], which includes the biologists most involved in protein crystallography, called in 1998 for the cancellation of the space-based program, stating that:

"No serious contributions to knowledge of protein structure or to drug discovery or design have yet been made in space." (ASCB, July 9, 1998)

Hoping to regain some credibility, an embarrassed NASA turned to the National Academy of Sciences to review biotechnology plans for the Space Station. On March 1, 2000, the National Research Council [NRC], the research arm of the Academy, released their study. It concluded that:

"The enormous investment in protein crystal growth on the shuttle and Mir has not led to a single unique scientific result." (NRC, 1 March 2000)

It might be supposed that at this point programs in space-grown protein crystals would be terminated. It was a shock to open the press kit for STS-107 [name for *Columbia* mission] following the *Columbia* accident, and discover that the final flight of *Columbia* carried a commercial protein crystal growth experiment for the Center for Biophysical Science and Engineering, University of Alabama at Birmingham. The Director of the Center is Lawrence J. DeLucas, O.D., Ph.D. If I go to the NASA web site and look for research planned for the ISS, I once again find protein crystal growth under the direction of the Center for Biophysical Science and Engineering and Dr. Lawrence J. DeLucas.

Microgravity Is Deleterious to Human Health

The microgravity environment has been found to be far more deleterious to human health than anyone had suspected. Indeed, in the first heady early days of the space age there was speculation that someday heart patients might be sent into orbit to rest their hearts, which would not have to pump blood against the force of gravity. On the contrary we find that not only is the heart severely stressed in zero gravity, osteoporosis, muscle atrophy, immune suppression, sleep disorders, diarrhea and bouts of depression and anxiety are endemic to the space environment.

By now you have all probably seen the "White Paper" by Dr. Lawrence Kuznetz that critiques the human life-sciences research aboard the ISS and the shuttle. Intended as an internal critique for his colleagues, the paper was leaked to the public. Kuznetz, a professor at the Baylor College of Medicine, and flight projects research manager for a NASA academic consortium, finds that few if any of the experiments have valid controls.

"The line between real and wishful science," he writes, "is continually being blurred." He puts the blame directly on NASA management. The stated objective of the life sciences research planned for the ISS is to develop "countermeasures" for the staggering number of health risks facing astronauts, particularly those who might someday venture beyond the relative safety of low-Earth orbit. "Under the worst of circumstances," he writes, "ISS will be in the ocean without a single countermeasure in the books for the cardiovascular, neurovestibular, pharmacokinetics, behavior and other major disciplines. Then again, we could get lucky."

It is unfortunate that in our democracy, conscientious public servants, willing to risk their careers by leaking documents to the public, may be the only protection we have against self-serving and misleading public pronouncements by government agencies. What's behind this is the NASA conviction that the public will not support a space program that does not involve putting humans in space. Research planned for the ISS is merely an extension of the sort of science conducted on the space shuttle over the past 20 years.

The Research Is Not Very Important

The research is not wrong, it is just not very important. No field of science has been significantly affected by research carried out on the shuttle or on Mir at great cost. Much of it has never even been published in leading peer-reviewed journals.

The real objective of the most expensive science laboratory ever constructed is to provide astronauts with something to do. Ned Ludd, an English laborer who destroyed weaving machinery in 1779 to preserve jobs, would have cheered. But human progress is now measured by the extent to which machines are used to replace humans to perform tasks that are dangerous or menial.

Even if shielding is added to spacecraft to protect against radiation, and a long axis spacecraft is rotated to provide artificial gravity at great cost, the only conceivable new destination for human explorers is Mars. Conditions on other planets or their moons are too extreme for humans to ever set foot on them. They are too hot, or their gravity would crush a human, or radiation levels are much too intense.

Robots, Not Humans, Are Needed

Mars is no garden of Eden either, but the 1997 *Pathfinder* mission to Mars gave us a glimpse of the future. *Pathfinder* landed on Mars carrying a lap-sized robot named *Sojourner.* The tiny robot caught the imagination of people everywhere. *Sojourner* was a telerobot. Its brain was the brain of its human operator 100 million miles away on Earth. Its senses were the senses humans gave it. The whole world saw Mars through *Sojourner*'s eyes. It had an atomic spectrometer for a nose that could sniff the rocks to see what they were made of, and thermocouples that could feel the warmth of the midday sun in the sand beneath its wheels. It never stopped for lunch or complained about the cold nights. Trapped in their space suits, human explorers could have done no more. Two much more sophisticated telerobots are now on their way to Mars. [They landed January 2004.]

Meanwhile, the exploration of space can't wait for astronauts. Our robots have already visited every planet save distant Pluto, testing the Martian soil for traces of life, and mapping the hidden surface of the cloud-shrouded planet Venus with radar eyes. Long before a human expedition to Mars could be launched, the robots will have finished their exploration.

We must ask what it means to "be there." Telerobots are robust extensions of their frail human operators, giving us a virtual presence in places no human could ever venture. The accomplishments of the astronauts on the ISS will be inconsequential. It is the scientists who control the telerobots, having become virtual astronauts, who will explore the universe. To explore where no human can ever set foot is the great adventure of our time.

NASA's Projects Are Expensive, Wasteful, and Inefficient

By Josh London

According to Josh London, NASA is a massive bureaucracy that over-spends, has poor programs, and survives by offering good deals to Congress. NASA is conducting worthless scientific experiments, London says in the following article, including one to study the existence of global warming, which NASA's satellite data have consistently shown does not exist. In addition, London says that the shuttle program and the International Space Station (ISS) are examples of NASA's wasteful and inefficient spending. NASA introduced the shuttle program with its reusable space vehicles, as a money-saving plan, but in reality, London says, NASA's spending has increased. Because the shuttle program was a commercial failure, London says, NASA had to justify itself by promoting a new International Space Station. The ISS was to cost $8 billion and be launched by 1992, but after eight years and $10 billion in expenditures, NASA produced only plans for the space station, resulting in the space station's near termination by Congress. In addition, London believes, NASA is sacrificing other scientific projects to support the ISS program. Josh London works as deputy editor for the *American Spectator.*

The laws of government, like the laws of physics, apply everywhere—whether on earth or in the dark reaches of outer space. When governmental bodies are funded without strict measurable objectives and deadlines, the result will be waste,

Josh London, "Pigs in Space," *The American Spectator*, vol. 33, November 2000, p. 34. Copyright © 2000 by *The American Spectator*. Reproduced by permission.

pork, corruption, and incompetence.

So it is with NASA. When President [Dwight] Eisenhower created the National Aeronautics and Space Administration on July 29, 1958, it was a shining testament to the power and grandeur of big government in its infancy. The space agency's first "big hit" was in February 1962 when John Glenn became the first American to orbit the earth. Its next big success, indeed NASA's crowning achievement, was the first manned lunar landing on July 20, 1969.

In the 31 years since then, NASA has made fantastic progress. Indeed, its most recent achievement was . . . Well, it was October 29, 1998, when John Glenn was shot back into space. Hmm . . . Something seems amiss.

Outer space is a place of numerical extremes. The distances it spans and the amounts of money required to explore it are both measured in digits that make the eyes spin like a slot machine. But while the universe expands at a steady, predictable rate, space budgets have an unnerving tendency to inflate suddenly and dramatically.

NASA Is a Federal Bureaucracy

The answer lies in the simple, commonplace, and lamentable fact that NASA is today as much a federal bureaucracy as the department of Housing and Urban Development (HUD). True, there are some basic differences. Most NASA employees can read, write, and do sums rather well; NASA is somewhat smaller, and HUD isn't rocket science. But NASA's primary unofficial purpose, like all federal bodies, is to keep its staff employed. And, like all federal bodies, NASA survives by offering valuable pork to Congress—considering that it started with rocket launches in one state (Florida) and "mission control" in another (Texas), NASA was seemingly created in a pork-barrel. NASA's two most significant projects—the shuttle and the space station—serve to illustrate.

The Space Shuttle

In 1972, as the *Apollo* program came to a close, NASA sold the shuttle program as a way to establish routine, low-cost, reliable access to space. Instead of Expendable Launch Vehicles that burned up after take off, Reusable Launch Vehicles could be used again and again, and save money.

Thanks to the "cost saving" Reusable Launch Vehicles, the price tag of hefting a pound of payload into space has increased from $3,800 in the 1960's to $6,000 (in constant dollars). And this is only an estimate. Dr. Alex Rowland, formerly a NASA historian but now a history professor at Duke University, calculates that, once the development and capital costs of the shuttle are factored in, the actual per-pound cost is $35,000. In other words, the cost of one shuttle flight is not $350 million as NASA claims, but much closer to $2 billion.

The International Space Station

As it became increasingly obvious that the shuttle program was a commercial bust, NASA needed a mission to justify the shuttle's—and not inconsequentially their own—continued existence. Thus, the quest for a new space station.

Initially named "Freedom," the American space station was to measure 500 feet, cost a mere $8 billion, and be launched and ready for activity by 1992. So over the next eight years NASA went to work, spending $10 billion and producing . . . nothing. Nothing but plans.

Even Congress couldn't ignore such waste—at first—and the project was slated for termination. But NASA was handed a gift: the collapse of the Soviet Union. Concerned that the defense industry would wither in recession after post–Cold War cutbacks, Congress turned to the International Space Station as the aeronautics contractors' way station.

In 1993, President [William] Clinton ordered NASA to design a less expensive station that could include Russia as a cost-sharing partner. The make-work nature of the decision was so transparent that it barely escaped the gravity of congressional objection. In June 1993 a proposal to cut off funding for the station lost by only one vote: 216 to 215.

The Space Station Is Just a Jobs Program

So NASA did what any good government agency would do: It sat down to reinvent and restructure, from head to toe, its lobbying efforts. It promised to create a "procurement constituency" of 40 states. In effect, NASA offered Congress a jobs and pork bonanza. As Albert Wheelon, a former executive at Hughes Aircraft and member of the 1993 presidential space station commission, summed it up: "The [space station] is now just a jobs program."

Florida, Texas, and Alabama were already locked in because much of the station work would be done within their borders. Not surprisingly the congressional delegations of all three states were unanimous in their support. The station's primary contractor is Boeing, with offices in Seattle, Houston, and Huntsville, Alabama. In recent years, however, NASA has spread the money to 67 other prime contractors and 35 major subcontractors in 22 states. The pork was sufficiently succulent to change congressional minds: In 1994, the space station's one-vote margin became a hefty 123-vote margin. By 1999, the margin of victory was secure: 245 votes for pigs in space.

The Clinton administration promised that U.S. spending—not including the $10 billion that was thrown away during the first eight years of the space station project—would not exceed $17.4 billion. But include the cost of Russian holdups (repairing and updating aging equipment) and bailouts (much of which the Russians cannot account for) and the cost comes closer to $25 billion. And that doesn't include the additional $15 billion to $20 billion for shuttle flights to ferry the station's parts out to space.

Add the expected two decades of expensive consulting and repair work and the total price tag for the U.S. component of the International Space Station becomes upwards of $96 billion, says the General Accounting Office.

Allowing the pedestrian facts of economics and politics to hinder space exploration, however, tends to ruffle feathers at NASA. The pro-NASA response-patter is bound tightly with notions of real scientific and technological progress. Through NASA, it is said, the nation can invest in its future.

The Space Station Is Scientifically Unnecessary

Yet NASA isn't exactly a stepping stone to technological advancement. Already NASA has sacrificed planned scientific shuttle and satellite projects in order to throw more resources at the International Space Station—a triumph of public relations and politics over hard science, particularly since the space station is so scientifically unnecessary. NASA even sacrificed planned biological experiments on the station itself to compensate for cost overruns in building the station. Clearly the pursuit of knowledge is slipping down this project's priority list.

Which is one reason why many American scientists oppose the space station. Another reason is that space research itself is far more cost effective when done by satellites and unmanned probes—most of which can be launched for the price of a single space-station resupply mission (and that's just using NASA's figures). About the only research the International Space Station will excel in is long-term observation of the physiological effects of space. And we already have much data on this from Russia's Mir space station—though even if we didn't one wonders if it would be worth close to $100 billion. As the special presidential advisory commission set to evaluate NASA's space station plan stated in 1991, "We do not believe that the space station . . . can be justified solely on the basis of the [nonbiological] science it can perform, much of which can be conducted on Earth or by unmanned robots."

Some of NASA's Experiments Are Worthless

The chairman of the commission neglected to mention that some of NASA's experimenting in space is fairly worthless science to begin with: What does it matter whether spiders spin their webs differently in zero gravity? What value is there in redundant zero-gravity fluid control experiments? Why spend hundreds of millions of dollars discovering the effects of weightlessness on an aged John Glenn?

Or take the "Mission to Planet Earth." MTPE was initially established by President [George H.W.] Bush in an effort to appease environmentalists. Its goal, as expanded by Vice President Al Gore, is to prove the existence of global warming by having a dozen satellites hovering around the planet taking the earth's temperature. All this despite the fact that NASA's satellite data has consistently demonstrated that the phenomenon does not exist. Project cost: $33 billion over 25 years.

Typical of NASA's scientific experiments is the gallium arsenide (GaAs) semiconductor wafer project. The GaAs wafer is used in producing the GaAs computer chip. NASA was convinced that a superior GaAs wafer could be produced in space—the zero-gravity would promote a nearly perfect crystal structure. So NASA spent a whopping $2.5 billion on five shuttle flights to make five GaAs wafers. The cost was about $100 million per wafer versus the $500 per GaAs wafer cost of the conventional terrestrial variety—not much of a commercial improvement. According to T.J. Rodgers, CEO of Cypress Semiconductor (the largest GaAs chip maker in the country), since the GaAs computer chip manufacturing process entails destroying and rebuilding the GaAs crystal, NASA's experiment was based on an entirely asinine hypothesis. It was, as Rodgers put it, "simply ludicrous."

The quest for knowledge will always challenge man's material and intellectual resources. But until there is a return on our investments into NASA, space exploration won't produce any new knowledge, only expensive demonstrations of old truths: Federal bureaucracies swallow taxpayer money whole. In this respect, NASA is better than federal agencies like HUD: At least NASA, at the end of the day, shoots the feeble, feckless fruits of its labor into space.

NASA Has Made Important Technological and Economic Contributions

By Todd Wilkinson

Scientists at NASA have been responsible for some of the most revolutionary advancements in space and aviation technology. In the following selection Todd Wilkinson explains that scientists and researchers at the little known NASA center in Cleveland, Ohio, named in honor of the pioneering astronaut John Glenn, played a pivotal role in the development of rocket propulsion in the race to space. Wilkinson reports that NASA Glenn's investments in technology have yielded billions of dollars in benefits for the U.S. economy by creating jobs and spin-off technologies and have channeled research-driven technology into U.S. industry. This has enabled major advances in commercial products like jet engines and communication satellites. In addition, NASA Glenn scientists oversee the development of the International Space Station as well as conduct research in biomedicine. For example, researchers are designing a helmet with eye-examining goggles to detect health abnormalities not only for space explorers, but for those in remote locations on Earth where medicine is unavailable. Todd Wilkinson is a writer, columnist, and western correspondent for several magazines, including the *Denver Post*, the *Christian Science Monitor*, and *Montana Magazine*.

On airy moonlit nights, stargazers in the Northern Hemisphere may notice what appears to be a glowing white speck making regular passes through the sky. It's not a UFO they are seeing or even the pulses of a meteor shower. That piece of metallic glitter is actually a massive human stepping-stone to the cosmos—the new International Space Station—orbiting 220 miles above the earth and taking shape as a base camp for the future exploration of our solar system.

Back on the ground, scientists and biomedical researchers from the National Aeronautics and Space Administration (NASA) are paying special attention to the space station's evolving construction from laboratories located in Cleveland. That's right, Cleveland. As in Ohio. The city pressed up against the southern shore of Lake Erie.

NASA Glenn

Surprising to many is that quietly over the past half-century some of the most revolutionary advancements in space and aviation technology have been developed at Lewis Field. The Glenn Research Center here, named in honor of the pioneering astronaut and U.S. senator, John Glenn, is perhaps the most unsung of NASA's 10 major campuses. Less known than the Johnson Space Center in Houston or the Kennedy launch pads at Cape Canaveral, Fla., or the Jet Propulsion Laboratory in Pasadena, Calif., NASA Glenn is, nonetheless, playing a pivotal role in transforming the agency's 11th and most novel facility—the space station—from a pie-in-the-sky dream into a symbol of 21st-century ingenuity. And it is giving Cleveland and numerous partner businesses and local universities a tangible connection to the frontier of space.

The NASA Glenn Campus is a labyrinth of six wind tunnels and more than 150 buildings, along with a beehive of laboratories. Since the early 1940s, around the time America entered World War II, the research facilities have been central to the development of jet engines that are today the foundation of commercial and military aviation. But in 1961, when President John F. Kennedy set U.S. sights on the moon, the laboratories also became nurseries for rocket propulsion in the race to space, notes Donald Campbell, director of the Glenn Research Center.

Better than any political leader in the country, Senator Glenn has understood the dividends accrued from public investment in technology. During recent heated debates in Congress over funding for NASA and concerns about cost overruns that have dogged the space station, it was Glenn who urged colleagues to support research and development in emerging technologies. If the United States is to maintain a competitive edge over other nations, he argued, it must sustain and nurture institutions like NASA.

NASA Technology Helps the U.S. Economy

Campbell says NASA Glenn channels much of its research-driven technology into U.S. industry, enabling major advances in commercial products like jet engines and communications satellites. During the 1970s and 1980s, NASA spent about $200 million on turbine engine technologies developed by Glenn and its commercial partners. In turn, that investment yielded billions of dollars in benefits for the U.S. economy, through job creation and spin-off technologies, including the eventual production of the General Electric 90 engine—the workhorse of many planes. "Engine propulsion technology has historically led the development of new generations of aircraft design, and that shows no signs of changing," says Joe Shaw, chief of NASA Glenn's ultraefficient engine technology program. "More and more we are seeing a cross pollination of ideas between the dual missions of NASA—its support of aeronautics for commercial and military purposes and exploration of space."

Likewise, the quest to build more powerful and efficient spacecraft reaped incredible dividends. "It's hard to tell what could come out of our space research that will affect our lives on the ground," Shaw says. "I don't think anybody with the *Apollo* program knew it would lead to the proliferation of personal laptop computers and digital wristwatches and microbiological sensors."

Dramatic Advancements in Airplanes

Not far off on the horizon, Shaw says, are aircraft that will burn dramatically cleaner fuel, reducing carbon dioxide and nitrogen oxide emissions that contribute to global warming and smog. Those same planes will boast engines that are barely audible to the human ear on the ground once the planes are beyond airport boundaries. Yet the biggest advancement that could arrive in less than a generation will be fleets of "smart airplanes," whose computer systems adjust engines in flight to make them fly more efficiently. And where commercial flights are concerned, efficiency results in the need for less fuel. Ultimately, that would mean better bargains for travelers. An ambitious goal of NASA Glenn scientists is to reduce the travel time to the Far East and Europe by half within the next 25 years, but to also make it possible at today's ticket prices.

Last September, *R&D* [*Research & Development*] Magazine named three research teams based at Glenn winners of its prestigious *R&D* 100 Award, known within the industry as the "Nobel Prize of applied research." The projects that attracted global attention involved the development of superstrong titanium alumnide sheet metal used in aircraft bodies; advancements with PMR (Polymerization of Monomer Reactants) to give aircraft longer shelf lives; and the application of GENOA software that has enabled Boeing and GE aircraft engines to save millions of dollars improving the cutting-edge 777 aircraft engine. Since the early 1960s, Glenn researchers have claimed nearly 80 of the 110 *R&D* 100 Awards given to NASA projects.

Awe-Inspiring Projects in Space Travel

Without question, the most awe-inspiring projects are those dealing with space travel. By his own admission, John Dunning, a 30-year NASA veteran and manager of space station support at Glenn, isn't a man prone to spontaneous gleeful outbursts. But last November [2000], when Space Shuttle *Endeavor* lifted off from the launch pad at Kennedy Space Center, Dunning and his Glenn colleagues let out a collective whoop. In her belly, *Endeavor* carried solar panel arrays and advanced nickel-hydrogen batteries that are today providing the power essential to making the International Space Station operational. Without the electrical juice generated by the photovoltaic panels and stored in super batteries, astronauts would be whistling in the dark, says Dunning.

Much of the transportable power grid, built and tested in cooperation with a handful of private aerospace companies, originated on drawing boards at the Glenn laboratories. Prior to shuttle launches in October, November [2000] and January [2001] a specially designed radiator that removes waste heat from the station was tested in the Space Power Facility, the world's largest space environment simulation chamber, at NASA Glenn's Plum Brook

Station in Sandusky, Ohio. "Before these recent shuttle missions delivered the power components, the space station crew had been confined to a service module, because most of the structure was uninhabitable," Dunning says. "With the power systems up and running, the volume of space available to crews will significantly improve by about a factor of three, and the amount of consumable electricity will increase from four kilowatts to 24 kilowatts."

Revolutionary Energy Storage Technology

A future principal component of the station's power plant, being developed by NASA Glenn, could be the "flywheel energy storage system," which functions like a gyroscope motor spinning at 60,000 revolutions per minute. When the space station arrays are illuminated by the sun, the flywheel functions like a mechanical battery, converting motion into usable energy and vice versa. During periods of orbit when the station is shaded from sunlight, the wheel is turned into a generator that makes electricity to power the life support system and science equipment. Scientists note that at full operating speed the flywheel rotor's linear velocity is two-and-one-half times the speed of sound (1,875 miles per hour). If the wheel itself were allowed to spin without meeting resistance, it would go on for more than 12 hours.

"The flywheel energy storage system represents a revolutionary step in energy storage technology," says Raymond Beach, NASA Glenn's team leader for flywheel development. He sees the flywheel as a potential long-term alternative for chemical batteries, which don't last as long and which generate waste. "The process is very efficient," he points out. "More than 85 percent of the energy put into the wheel comes out."

Solars Cells on Mars

NASA believes that in the coming decades similar solar-powered generators could have applications on earth and on Mars. When the Mars Surveyor Lander mission reaches the Red Planet, two pilot Glenn projects—the Mars Array Technology Experiment (MATE) and the Dust Accumulation and Removal Technology (DART)—will explore the feasibility of producing oxygen propellant from the Martian atmosphere and will test whether power-generating solar cells can function amid extreme cold and notorious Martian dust storms. "Because of the dust, the cold temperatures and the varying light spectrum, the best solar cell for our 'gas station on Mars' might be one that we wouldn't consider using in our space solar arrays," says NASA Glenn Project Manager Cosmo Baraona, who is overseeing the experiments.

Solar cells designed at Glenn have already performed better than expected with the *Pathfinder* and *Sojourner* Rover, but David Scheiman, a researcher at the Ohio Aerospace Institute in Cleveland, a partner of Glenn, says it is uncertain if those cells will work over the estimated five years it will take to get a human to and from Mars.

A Hub for Biomedicine

Through its Microgravity Science Division, Glenn is NASA's star performer with microgravity experiments involving combustion and fluid physics. Aside from its history with spacecraft and jet engines, Glenn has bolstered Cleveland's reputation as a hub for biomedicine. "We are fortunate to reside in a region with some of the best medical research institutions in the country and a growing biomedical industry base," says Campbell.

At the forefront are researchers like Rafat Ansari, a groundbreaking physicist. "My personal interest is with the human eye," he says. According to Ansari, our eyes are not only windows to the soul, but also windows to the human body, reflecting the health and function of vital chemical processes. They are also places where physicians can look to better understand the risks of exposure to radiation during deep space travel to destinations like Mars. "When light passes from the cornea into the retina, it also passes through nearly every tissue type found in the body," Ansari says. "By studying those tissues, we can look for evidence of certain conditions from one's cholesterol level to the formation of cataracts to the potential for Alzheimer's disease to diabetes."

Ansari began his career with NASA 13 years ago. His fascination with eyes started when his father developed cataracts. It led him to investigate the etiology of cataracts and the risks associated with certain diseases. Astronauts can be especially vulnerable because increased exposure to radiation associated with deep space travel may accelerate the growth of cataracts and macular [area in the retina] degeneration.

Ansari and a team of Glenn researchers are working with the

federal Food and Drug Administration to develop a screening process for diabetes. Another project at the Glenn laboratories involved development of an apparatus in partnership with the National Eye Institute, located at the National Institutes of Health in Bethesda, Md. It would have applications not only on Mars but also in rural parts of the world where there is a niche to fill with telemedicine. The patient or, in the case of space travel, the astronauts would wear a specially designed helmet with eye-examining goggles connected to special sensors monitoring the heart in real time. The apparatus could detect health abnormalities as explorers walk across the Martian surface. But long before the first human mission is sent to the fourth planet from the sun, Ansari would like to see such mobile devices used in remote locales on earth where medicine is unavailable.

In the years ahead, the facility bearing Senator Glenn's name promises to claim its own prominent place on the journey of human discovery. "[In 2001] as we celebrate the Glenn center's 60th anniversary, all of us can look back in pride at our outstanding accomplishments that have helped propel NASA and U.S. industry to new horizons," adds Campbell. "And no matter where that next horizon is found, Glenn's pioneers and innovators will make it possible for us to travel beyond it. Ultimately, we want the public to benefit from what we do."

Space Exploration Should Be Supported by a Space Tourism Business

By Buzz Aldrin

The United States must be committed to human exploration in space as well as to a permanent and profitable presence there. Former astronaut Buzz Aldrin expressed these beliefs at a hearing before the Subcommittee on Space and Aeronautics in 2001. Aldrin advocated the creation of a thriving tourism industry that would shuttle citizens to Earth-orbiting hotels. He also promoted the settlement of the moon in 2015 and of Mars in 2020. Aldrin envisioned Earth, moon, and Mars as busy hubs for the ebb and flow of passengers, cargo, and commerce. He believed these goals were practical, achievable, and affordable, based on the space program's decades of expertise in the shuttle and space station programs.

Buzz Aldrin was the lunar module pilot for the *Apollo 11* mission on July 20, 1969, the first manned-mission to the moon, and was the second human to step onto the moon. After retiring from NASA, Aldrin founded the Starcraft Boosters, a private rocket design firm, and authored an autobiography, *Return to Earth*, as well as two novels, *Encounter with Tiber* and *The Return*.

Buzz Aldrin, address before the U.S. House Subcommittee on Space and Aeronautics, Committee on Science, Washington, DC, April 3, 2001.

orty years ago this coming April 12th [2001], Yuri Gargarin became the first human to see the Earth from space, an event that sparked President [John F.] Kennedy's commitment to put an American on the moon. Next month [May 2001] will mark the fortieth anniversary of that historic speech.

Kennedy belonged to the so-called "Greatest Generation"— people who were willing to accept risk and sacrifice, who had a vision of something larger than themselves, who abided depression and war and left America a colossus astride the Earth. As their last great gesture they put humanity on the moon.

One thinks of the dying lieutenant's last two words to Private Ryan: "Earn this" [in the movie *Saving Private Ryan*]. In our attempts to create a risk-free society, we've often failed to honor that debt. There is a failure of nerve in postmodern society. We seem to have reached a crossroads similar to that of sixteenth-century Europe on the eve of expansion into the New World—a crisis now more ominous than the cold war threat that compelled Kennedy's commitment. On the one hand, there is a loss of vigor, a spreading irrationalism, and a collective hypochondria that seems to cripple our larger visions. Funding for basic research and development continues to decline, while the dream of space exploration succumbs to the dream of animal comfort. "Where there is no vision," says the proverb of Solomon, "the people perish."

A New Renaissance

On the other hand, we stand at the threshold of a new Renaissance, a moment much like the morning of the modern age when most of the globe lay deep in mystery. What a time this is to be alive! Today's young will live to see settlements in space, unlimited energy from fusion, and explosions of knowledge on all frontiers— from the workings of the brain to the origin and nature of the cosmos itself.

The only obstacles to that future are complacency and a lack of clear commitment. If we insist that the human quest await the healing of every sore on the body politic, we condemn ourselves to stagnation. In the long run, the whole politics of society is more profoundly changed by a new sense of human potential than by any amount of obsessive self-maintenance. Like all living systems, cultures cannot remain static; they evolve or decline. They explore or expire. They take risks.

Risk has always been the price of any successful venture—whether it be our migration out of Africa into the northern ice, the discovery of the New World, the shaping of a continent, or the preservation of that new freedom. The continued exploration of the solar system is a challenge that can bind together nations, inspire youth, advance science, and ultimately end our confinement to one vulnerable world. Beyond all the political and economic rationales, momentous as they are, spaceflight is a spiritual quest in the broadest sense, one promising a revitalization of humanity and a rebirth of hope no less profound than the great opening out of mind and spirit at the dawn of the modern age.

The *Apollo* program showed not only what humanity can achieve with strong leadership and solid commitment, but also the capacity of such pursuits to arouse public participation and inspire a sense of purpose. Three decades after the event, people still feel compelled to tell me exactly where they were at the moment I walked on the moon. Yet history will remember the inhabitants of the last century as the people who went from Kitty Hawk to the moon in 66 years—only to languish for the next 30 in low-Earth orbit.

An Open Frontier for All People

If we are to resurrect the profound feeling of participation that accompanied *Apollo* we will need a Kennedy-like commitment to human exploration, which must begin with a permanent and profitable presence in space. This is why I've been intensely involved in an effort to put citizens on the Shuttle by lottery, and to develop cost-effective, reusable boosters to take tourists into space and foster the birth of an expanded "hospitality" industry in orbit. A new generation of space vehicles can carry private citizens to orbiting hotels, settlers to the moon and Mars, and waves of explorers to the far reaches of the solar system.

Beyond robotics and Earth-serving space stations lies the infinite journey. But within two or three decades, space can be an open frontier for all people. I see a near-term future where economical, two-stage space launchers place paying passengers and cargo into Earth orbit with the efficiency and routine-like nature of today's airline traffic. A booming tourism industry will be cultivated as space hotels become a point-of-arrival and departure above our planet. This burgeoning business enterprise will bring about heavy-lift rockets enabling grander civil steps of exploration, back to the moon, to the distant dunes of Mars, and beyond.

I further envision long-haul transportation systems, deep space cruisers that not only continuously cycle tourists between the Earth and moon, but constantly transfer explorers and settlers between Mars and the Earth. A fully reusable lunar and interplanetary system is the ultimate way of transporting people and cargo across the vast vacuum void of space.

An Action Plan for the Future

But how do we get there from here? I see an action plan for the future—a plan based on years of training and experience this country so graciously invested in me.

As our next step in space, lowering the cost of space access with a reusable two-stage-to-orbit launcher is critical. The first step is the incorporation of a Reusable First Stage into our space architecture. Sized properly, it will be a commercially competitive workhorse. It will hurl another rocket-powered vehicle, with payload, allowing it to reach space with greater economy than if purely self-propelled. By dropping the expense of attaining Earth orbit, many new industries are waiting to develop, one of which will be space tourism. Soon, tens of thousands of citizens will have the opportunity to travel into space, gaining a sense of "participation" in opening the frontier of space to enterprise, exploration and settlement.

From this step, an add-on to the reusable space program philosophy is building a "bridge between worlds." Through a system of reusable spacecraft that I call "Cyclers," traffic routes—first between Earth and the moon, then Mars and Earth—should be put in motion. Very much like ocean liners, the Cycler system would perpetually glide along predictable pathways, moving people, equipment, and other materials to and from the Earth over inner–Solar System mileage. A sequential buildup of a Full Cycling Network could be in place within two decades of a go-ahead, geared to the maturation of lunar and Mars activities. The Earth, the moon, and Mars will form a celestial triad of worlds—busy hubs for the ebb and flow of passengers, cargo and commerce traversing the inner–Solar System.

My schedule for accomplishing these objectives is practical, achievable and affordable, drawing from decades of space expertise already honed by our early exploits, including the Space Shuttle and International Space Station projects.

I call for a strong and vibrant space tourism business and a return to the moon by 2015, then reaching Mars by 2020. The common link between steps in this timetable is a progressive set of reusable boosters, reusable access to space, then reusable interplanetary Cyclers.

A New Commitment to Space

This vision spans two decades of enterprise, exploration and settlement. Ideally, it should be enunciated by the new U.S. President on the upcoming fortieth anniversary of President Kennedy's space commitment speech. By the year 2030, I see the same people looking back and cherishing the moment that a leader of our country committed us to a gradual, but progressive plan of permanent settlement of space, not just occasional visits that leave little more than flags and footprints.

The surface area of Mars is equivalent to the land area of Earth. Once a human presence on this planet is established, a second home for Humankind is possible. A growing settlement on Mars is, in essence, an "assurance" policy. Not only is the survival of the human race then assured, but the ability to reach from Mars into the resource-rich bounty of the Martian satellites and the nearby asteroids is also possible. These invaluable resources can be tapped to sustain increasing numbers of Martian settlers, as well as foster expanded interplanetary commerce and large-scale industrial activities to benefit the home planet—Earth. Of course, some will insist on building outer–Solar System Cyclers as humanity continues outbound into the Universe at large.

My vision is a call for a sustained space program. We can now chart a course that returns us to the moon, then allows humanity to strike out for the New World of our future—the planet Mars. But our near-term space efforts, both manned and robotic missions, must be tailored to support the longer-range purpose of opening the frontier. Step by step, program by program, we can construct a future of limitless potential. I must ask you gentlemen, if not for these bold endeavors, then what is our space program for?

Steps for Congress

Please allow me to address five significant points that Congress, as a governing oversight body, can do to help guarantee the vision I have described here today to come to fruition for future generations.

- First, the highest priority of NASA and Congressional guidance of NASA's activities must be to develop lower cost to orbit systems. Congress should continue its leadership role in this direction by expanding the spectrum of development options to now include two-stage-to-orbit systems beginning with a rationally sized, commercially competitive, Reusable First Stage vehicle.
- Second, continue to identify and eliminate stifling regulations that inhibit the private sector from competing in the commercial launch vehicle market to facilitate the development of lower cost space transportation options.
- Third, charge NASA with investigating an expanded scope of lower cost space transportation system options to include those options that promise to reduce costs by factors of two or three rather than to focus exclusively on only order of magnitude cost reduction improvements.
- Fourth, focus NASA and the private sector on the near-term objective of flying "people" in space and thoroughly assess the impact of flying tens of thousands of prospective paying passengers on the development and evolution of next generation of space transportation systems and new orbital industries.
- And finally, charge NASA to study in depth these recommendations including the reusable cycling transportation system I have described for economical exploration and development of the moon and Mars.

Space Should Be Protected as a Wilderness

By Ryder W. Miller

In the following selection Ryder W. Miller argues that those involved in space exploration need to be more conscientious about the environmental problems of space. In order to ensure environmental awareness about space, Miller suggests that the United Nations and NASA implement astroenvironmentalism. Astroenvironmentalism, Miller explains, is the application of environmental and preservationist values to developments in space exploration, militarization, and commercialism. Adopting astroenvironmentalism would ensure an adherence to a code of ethics that would help avoid environmental disasters in space such as those that have already occurred on Earth. It would also define space as a wilderness to be preserved rather than a frontier to be exploited. This would keep the space surrounding Earth free from pollution, debris, and garbage, as well as prohibit the national, international, or private ownership of property in space. Miller believes that discouraging property ownership in space would help avoid military conflict, the most important issue involved with the space environment. Miller says that space must not be a new battleground and that nuclear power should be kept from being used for military purposes in space.

Ryder W. Miller is a journalist who has published articles in *Mercury* magazine, *Space Policy Digest, Ad Astra* magazine, New Mars.com, and Gyre.org. He is a member of the Astronomical Society of the Pacific and editor of *From Narnia to a Space Odyssey: The War of Ideas Between Arthur C. Clarke and C.S. Lewis.*

Astroenvironmentalism, an argument to apply the values of environmentalism and preservationism to developments in space exploration, militarization and commercialism, is not a new idea. But recent developments in space exploration suggest this perspective is not widely acknowledged enough by those who envision taking steps to enter space. Environmentalists did not take a stand on these issues over the last few years, which was unfortunate because this was a topical time to argue that space should be an environmental issue. Astroenvironmentalism is an addition to present efforts, but also an umbrella term to describe a variety of related concerns held by many players in the environmental arena. Since mankind made such a mess of this planet and is now paying the environmental price for the damage, this topic is of extreme importance because we must avoid making the same mistakes in space as we have on Earth. At issue are the environmental consequences of the steps we are about to take in entering space. The adaptation of environmental concerns to developments in the exploration and commercialization of space fit surprisingly easily. Astroenvironmentalism is another reformulation of the associated environmental concerns involving a space wilderness to protect, rather than a "frontier" to exploit.

Some Concerns of Astroenvironmentalism

As I have outlined elsewhere, some of the concerns of astroenvironmentalism can include:
- Keeping the space surrounding the Earth clear of pollution, debris, and garbage. Efforts are necessary so we do not add to the reservoir of human waste and machinery left behind by space explorers. Such debris could cause damage to satellites and the space shuttles.
- Remembering and teaching the lessons learned from terrestrial conservation and preservation struggles of the past and applying them to the new frontier of space, that is, considering space and the celestial bodies pristine wildernesses that need to be protected rather than frontiers to conquer.
- Tracking and monitoring the environmental damage caused by the fuels used for space expeditions, that is, making space agencies adhere to the restrictions of environmental impact statements. In particular, it would be worthwhile to reduce the amount of plutonium that is being used in case of a mishap

that would result in plutonium entering the atmosphere.

- Treating the moon, Mars, Venus, and other planetary bodies as wildernesses that need to be protected, that is, arguing against the idea to "terraform" these celestial bodies. Terraforming introduces atmosphere-creating life into the barren celestial bodies in the effort to make these celestial bodies more amenable to human settlement. Terraforming is presently being explored despite the fact that we have not thoroughly explored these planets for indigenous life.

- Creating a set of ethical guidelines to protect the life that we encounter elsewhere, that is, study and protect rather than just study. The creation or re-publicizing of ethics applied to these concerns would be welcome.

- Creating safeguards to insure there is no contamination of celestial bodies, that is, safeguarding against the introduction of non-terrestrial life to and from celestial bodies. Non-indigenous life, whether it be Zebra mussels or microbes, under conditions where there are no controlling factors, can reproduce at exponential rates thereby changing the environment in the process. These changes can harm the organisms that were dependent upon the original environmental conditions.

- Counteracting the efforts of national and private agencies to terraform other planets. This idea to terraform is not just science fiction, and ecocritics can criticize science fiction writers who want terraforming to occur before a thorough search for life is conducted. This has been evident in Kim Stanley Robinson's award-winning science fiction trilogy *Red Mars, Green Mars and Blue Mars*, and recent films such as *The Ghosts of Mars* and *Red Planet.*

- Prohibiting national, international, and private agencies from owning property in space, in the interest of avoiding military conflicts. There is a need for more people to be involved in the efforts to see that space does not become another battleground.

- Creating the legal power to enforce these concerns. This would make more people aware of international space law and the need to enforce it. The United Nations rules on such issues through the Committee on the Peaceful Uses of Outer Space.

The most important related efforts are those involved in trying to stop the militarization of space and the use of nuclear power in space. Karl Grossman, author of *The Wrong Stuff* (1997), and William E. Burrows, in *This New Ocean* (1998), point out that space is likely to become our next war zone. Space will become the new high ground from which battles are fought. We have ignored the moon so that we can focus more on the immediate high ground in the satellite belt. Thankfully, we are focusing on international cooperation for the new space station, but Grossman and Burrows emphasize the need for a greater worldwide participation.

Over the years there have been many people who have been concerned with this issue, but they would not necessarily call themselves astroenvironmentalists. I put forth astroenvironmentalism as an argument that space should be considered an environmental issue and the term can function as an umbrella term for the related concerns. Astroenvironmentalism seems to fill a void, because there are no widely known organizations that focus on this issue. There is no widely known Mars First or Venus First organization arguing against terraforming. There is no Greenspace or Spacepeace. Most environmental groups are focused on more immediate issues and are more concerned with immediate and down-to-Earth issues. Leopold's *Land Ethic*, which focused on protecting life, is not easily applicable to the barren territories of space. But the argument of protecting space from exploitation is not solely about protecting rocks; it is also about making a statement about human behavior. If one succeeds in making the argument about protecting celestial bodies, we are also making the argument about protecting habitats here on Earth.

In *Beyond Space Ship Earth: Environmental Ethics and the Solar System*, probably the most thorough coverage of the subject, [Eugene] Hargrove (1986) writes that the only reason there are no people on the moon or Mars is due to reduced NASA spending levels. "The attempts to apply environmental concepts to the Solar System represent a significant challenge for environmental ethics, since so far as we know at present the Solar System, except for Earth, is a collection of nonliving natural objects, the kind of entity that offers the greatest conceptual difficulties for environmental ethics." Hargrove warns, "If serious planning begins without adequate ethical and environmental input, then future NASA and associated industrial/commercial projects in the Solar System may simply produce a new environmental crisis that dwarfs our current one." Hargrove argues that if we do nothing, the dark visions of science fiction could become true. . . .

The Role of the United Nations

Many of these issues have been addressed by the United Nations Committee on the Peaceful Uses of Outerspace. It is illegal to claim property on the celestial bodies of the solar system, but what ability will the United Nations have to enforce such laws once there are private and national interests in space? The Committee has also mentioned applying environmental concerns to space, but when will they take action and how many qualified environmentalists are there to work on these issues? They have focused on space debris as a legitimate concern, but where will the funds to clean up space come from? Where will NASA find the funds to clean up the mess they have made?

The United Nations will need support if it is to achieve these goals and objectives. UNISPACE III, the Third United Nations Conference on the Exploration and Peaceful Uses of Outer Space, held in Vienna in July of 1999, included discussion of "solutions to problems of development" and the need for adherence to existing international space treaties and principles, but preservation of the solar system was only a minor focus of the proceedings.

The first goal of astroenvironmentalism should be to lobby the United Nations and NASA to require SpaceDev, The Artemis Society, The Mars Society, and the Astrobiology Institute [companies or organizations involved in various aspects of space exploration] to agree to ethical guidelines. What leaves the Earth reflects the whole species. These agencies should be required to agree to principles of preservation or astroenvironmentalism established by the United Nations and NASA. For the meantime they should publish environmental impact statements on the Internet for the world to read. If we only allow astroenvironmentalists into space then maybe we will not have environmental battles on other celestial bodies. If we do not pay attention, we are likely to face these problems again.

CHRONOLOGY

1957

October: The Soviet Union launches the first artificial satellite, *Sputnik I*, triggering the space race with the United States.
November: The Russians put the first animal into space, a dog named Laika, in *Sputnik 2*.

1958

January: The United States launches its first satellite, *Explorer I*. With this mission, scientists make the important discovery that radiation belts surround Earth.
October: The National Aeronautics and Space Administration (NASA) succeeds the National Advisory Committee for Aeronautics, which had been responsible for researching the aeronautical industry since 1915. NASA is now responsible for overseeing the civilian space program. Military space activities are conducted by the Department of Defense.

1961

April: Soviet cosmonaut Yuri Gagarin becomes the first human in space, inaugurating the era of manned spaceflights. The spacecraft carrying Gagarin, *Vostok 1*, makes one complete orbit around the earth. The flight lasts one hour and forty-eight minutes.
May: Alan Shepard becomes the first American in space when he conducts a fifteen-minute suborbital flight. President John F. Kennedy makes the now-famous speech that challenges the United States to land an astronaut on the moon before the end of the decade.

1962

February: John Glenn becomes the first American to orbit Earth.

1965

March: The first space walk is performed by Russian pilot

Alexei A. Leonov, who is tethered outside his spacecraft while in Earth orbit. This historic venture into space lasts twelve minutes.

1966

February: The unmanned Russian spacecraft *Luna 9* completes a 250,000-mile trip and makes a successful soft landing on the moon. *Luna 9* transmits pictures of the moon's surface back to Earth. The mission demonstrates that the moon's surface is strong enough to support the weight of a large spacecraft.

1967

January: In the first U.S. aeronautics tragedy, Project Apollo, the mission charged with reaching the moon, suffers a disastrous setback when a fire in the *Apollo 1* command module kills three astronauts. The disaster delays further manned flights for almost two years.

1968

September: The Soviet *Zond 5* is launched and becomes the first spacecraft to orbit the moon and return.

1969

July: Neil Armstrong and Edwin "Buzz" Aldrin of the *Apollo 11* mission become the first humans to walk on the moon.

1970

April: The *Apollo 13* spacecraft loses its main power supply due to an explosion on the way to the moon. To the relief of the astronauts aboard and the American public, NASA's ground controllers in Houston devise a way to use the moon's gravity field to shoot the spacecraft back to Earth.

1971

April: The USSR launches the first space station, *Salyut 1*. It remains in orbit until May 28, 1973.

1973

May: The United States launches *Skylab*, the nation's first space station.

1975

July: The United States and the Soviet Union cooperate on the *Apollo-Soyuz* test project, during which American and Soviet spacecraft dock in orbit, proving that the two countries can work together in space.

1977

August and September: The two historic *Voyager* missions leave Earth to fly by Jupiter in 1979 and Saturn in 1980.

1981

April: The first space shuttle, *Columbia*, takes off from Kennedy Space Center.

1983

June: Sally K. Ride becomes the first American woman to travel in space. Ride is on board one of the space shuttle *Challenger* missions.

1986

January: The space shuttle *Challenger* explodes just seconds after liftoff, killing all seven aboard, including teacher Christa McAuliffe, the first private citizen to participate in a spaceflight. This accident, which is blamed on design and program inadequacies, grounds the shuttle program for several years.
February: The Soviet Union launches *Mir* (meaning "Peace"), the first space station to be continually inhabited by rotating crews.

1990

April: NASA launches the Edwin P. Hubble Space Telescope. The device provides significant new information and discoveries about the universe, including astonishing images of deep space, supernovas, and black holes.

1993

August: The *Mars Observer* spacecraft disappears three days before it is supposed to enter orbit around Mars. The mission, which was to be the first U.S. mission to Mars since 1975, cost $900 million. Consequently, NASA expands efforts to create a less expensive, more expendable martian satellite and lander.

December: American astronauts perform a demanding series of space walks to make repairs to a flawed mirror on the Hubble Space Telescope. These repairs, generally regarded as the most challenging task ever performed in space, enable the Hubble to function at an optimal level of performance.

1995

June: The space shuttle *Atlantis* docks with the *Mir* space station. Astronauts are transferred to and from *Atlantis*. U.S. astronaut Shannon Lucid lives aboard the *Mir* for five months, logging more hours in orbit than any other female astronaut.

1996

November: The *Mars Global Surveyor* is launched, entering the martian orbit in 1998 (it has been mapping the planet ever since).

1997

February: A probe from the *Galileo* spacecraft reveals that Europa, a moon of Jupiter, may contain ice or liquid water. According to scientists, ice and water would indicate the possible presence of life-forms on the moon.

July: The *Pathfinder* spacecraft lands on Mars. The craft carries a miniature rover called *Sojourner* that explores the surface of the planet. *Pathfinder* transmits voluminous amounts of new information about Mars, including color images of the surface of the planet and analysis of martian rocks and soil.

1998

November: The first module of the International Space Station (ISS), a permanent, orbiting scientific laboratory where various research projects will be conducted, is launched into space. The United States, Canada, Russia, Japan, and eleven European nations sponsor the space station. The station is expected to be completed by 2006.

1999

May: The Hubble Telescope accomplishes one of its primary goals, to determine a value for the rate of expansion of the universe.

This rate, known as the Hubble constant, is a key to determining the age and size of the universe. The *Discovery* becomes the first shuttle to dock with the ISS when it delivers parts and supplies to the space station.

July: The space shuttle *Columbia*'s twenty-sixth flight, a mission to deploy the Chandra X-Ray Observatory, is led by Air Force Colonel Eileen Collins, the first woman to command and land a shuttle mission. Colonel Collins had previously piloted a shuttle in 1995.

November: Astronomers witness the shadow of a planet in orbit around a star. Astronomers had suspected the existence of planets beyond our solar system but had not been able to directly confirm their hypothesis until this sighting.

2000

November: An international crew begins living aboard the ISS. The American and Russian crew, commanded by American astronaut William Shepherd, live aboard the space station for four months.

2001

March: The abandoned Russian space station, *Mir*, reenters Earth's atmosphere in a controlled fall into the South Pacific after fourteen years in space.

April: American businessman Dennis Tito becomes the first tourist to fly into space. The United States had rejected his $20 million offer but it was accepted by the Russian space program. A *Soyuz* space capsule delivers Tito and a Russian crew to the International Space Station.

2003

February: The space shuttle *Columbia* explodes over Texas, shortly before its scheduled landing in Florida. The seven-person crew, six Americans and the first Israeli to go into space, is killed.

2004

January: NASA successfully lands two rovers, *Spirit* and *Opportunity*, on the surface of Mars and confirms the previous existence of water on the red planet.

FOR FURTHER RESEARCH

Books

Benjamin Adelman and Saul Adelman, *Bound for the Stars: An Enthusiastic Look at the Opportunities and Challenges Space Exploration Offers.* Upper Saddle River, NJ: Prentice-Hall, 1980.

Buzz Aldrin and Malcolm McConnell, *Men from Earth.* New York: Bantam, 1989.

Leland F. Belew, ed., *Skylab, Our First Space Station.* Washington, DC: National Aeronautics and Space Administration, 1977.

Laurence Bergreen, *Voyage to Mars.* New York: Riverhead, 2000.

Serge Brunier and Stephen Lyle, *Space Odyssey: The First Forty Years of Space Exploration.* Cambridge, UK: Cambridge University Press, 2002.

Andrew Chaikin, *A Man on the Moon.* New York: Viking, 1994.

Arthur C. Clarke, *The Exploration of Space.* New York: Pocket Books, 1979.

Michael Collins, *Liftoff: The Story of America's Adventure in Space.* New York: Grove, 1988.

Gene Farmer and Dora Jane Hamblin, *First on the Moon.* Boston: Little, Brown, 1970.

Timothy Ferris, *Life Beyond Earth.* New York: Simon and Schuster, 2000.

Sarah Flowers, *Space Exploration: A Pro/Con Issue.* Berkeley Heights, NJ: Enslow, 2000.

Tim Furniss, *The Atlas of Space Exploration.* New York: Friedman/Fairfax, 2002.

———, *The History of Space Vehicles.* San Diego: Thunder Bay, 2001.

Robert Jastrow, *Journey to the Stars: Space Exploration Tomorrow and Beyond.* London: Transworld, 1990.

Roger D. Launius, *Frontiers of Space Exploration.* Westport, CT: Greenwood, 1998.

Jerry M. Linenger, *Off the Planet.* New York: McGraw-Hill, 2000.

Christopher Mari, *Space Exploration.* New York: H.W. Wilson, 1999.

Michael M. Mirabito, *The Exploration of Outer Space with Cameras: A History of the NASA Unmanned Spacecraft Missions.* Jefferson, NC: McFarland, 1983.

Andrew Mishkin, *Sojourner: An Insider's View of the Mars Pathfinder Mission.* New York: Berkeley, 2003.

Bruce C. Murray, *Journey into Space: The First Three Decades of Space Exploration.* New York: W.W. Norton, 1989.

David Owen, *Into Outer Space: An Exploration of Man's Obsession and Interaction with the Cosmos: Fact and Fiction.* Chicago: McGraw-Hill/Contemporary, 2000.

Carolyn Collins Petersen and John C. Brandt, *Hubble Vision: Astronomy with the Hubble Space Telescope.* New York: Cambridge University Press, 1998.

Harry L. Shipman, *Space 2000.* New York: Plenum, 1987.

G. Harry Stine, *Living in Space: A Handbook for Work and Exploration Beyond the Earth's Atmosphere.* New York: M. Evans, 1997.

Carole Stott, *Eyewitness: Space Exploration.* New York: DK, 2000.

Richard Wagner and Howard Cook, *Designs on Space: Blueprints for 21st Century Space Exploration.* New York: Simon and Schuster, 2001.

Periodicals

Nick Allen, "Miracles on a Shoestring: Russia's Space Odyssey," *Russian Life*, September/October 2003.

Andre Balogh, "The Future of Space Exploration: Is It Asking for the Moon?" *Contemporary Physics*, September 2001.

Marina Benjamin, "The End of the Space Age," *New Statesman*, February 10, 2003.

Michael Carroll, "New Discoveries on the Horizon," *Astronomy*, November 1995.

Tim Cavanaugh, "Space Balls," *Reason*, April 2003.

Andrew Cheng, "Pluto or Bust," *Astronomy*, May 2002.

Geoff Chester, "Martian Chronicles," *Astronomy*, August 2003.

Arthur Clarke, "Beyond Gravity," *National Geographic*, January 2001.

Stefano Coledan, "Death of a Space Station," *Popular Mechanics*, June 2001.

Ron Cowen, "Captured on Camera: Are They Planets?" *Science News*, May 26, 2001.

Ian Crawford, "Where Are They?" *Scientific American*, July 2000.

Current Events, "Pathways to the Stars," December 1, 2000.

Tad Daley, "Our Mission on Mars," *Futurist*, September/October 2003.

Orville F. Desjarlais Jr., "Stepping into the Unknown," *Airman*, September 2003.

Nader Elhefnawy, "Beyond *Columbia*: Is There a Future for Humanity in Space?" *Humanist*, September/October 2003.

Rand H. Fisher and Kent B. Pelot, "The Navy Has a Stake in Space," *Proceedings of the United States Naval Institute*, October 2001.

Sam Flamsteed, "Lost: The Mars *Observer* Probe," *Discover*, January 1994.

Geology Today, "Extraterrestrial Watch," March/April 2000.

William Hartmann, "Welcome, Earthlings," *Maclean's*, February 9, 2004.

Edward Hutchings Jr., "The Autonomous *Viking*," *Science*, February 18, 1983.

Richard A. Kerr, "NASA's New Road to Faster, Cheaper, Better Exploration," *Science*, November 15, 2002.

Andrew Lawler, "Onward into Space," *Astronomy*, December 1998.

William Lowther, "Lost in Space: A Mission to Mars Turns into a Costly Fiasco," *Maclean's*, September 6, 1993.

David Malakoff, "A $100 Billion Orbiting Lab Takes Shape: What Will It Do?" *Science*, May 14, 1999.

Joe McNally and Lisa Sonne, "Space Man of the Year," *Life*, January 1998.

Richard Monastersky, "Mars Mania," *Chronicle of Higher Education*, January 23, 2004.

Scott Mowbray, "After *Columbia:* The ISS in Crisis," *Popular Science*, April 2003.

Stephen J. Mraz, "Candidates for Exploring Mars: Four Innovative Exploration Programs Vie for the Honor of Going to Mars," *Machine Design*, July 10, 2003.

Robert Naeye, "Blazing a Trail to the Red Planet," *Astronomy*, October 1997.

Mark Nichols, "1998: A Space Odyssey," *Maclean's*, November 30, 1998.

Charles W. Petit, "Fit and on a Mission," *U.S. News & World Report*, January 18, 2000.

————, "Happy Landing," *U.S. News & World Report*, January 19, 2004.

Richard R. Vondrak and Dan H. Crider, "Ice at the Lunar Poles," *American Scientist*, July/August 2003.

Ronald J. White and Maurice Averner, "Humans in Space," *Nature*, February 22, 2001.

Jim Wilson, "48 Hours in Mission Control," *Popular Mechanics*, October 1997.

David A. Wolf, "Wow! The Earth," *Whole Earth*, Winter 1999.

Colin Woodard, "Stuck in Orbit," *Bulletin of the Atomic Scientists*, March/April 2001.

Web Sites

NASA, www.spaceflight.nasa.gov/history/index.html. This is one of NASA's Web sites with information about the history of space missions, including Mercury, Gemini, Apollo, and space shuttle missions. It includes time lines and biographies for each launch program.

National Space Society (NSS), www.nss.org. The National Space Society envisions people living and working in communities beyond Earth. The NSS Web site offers space resources and links as well as information on legislative and grassroots space tourism.

Nauts.com, www.nauts.com. Nauts.com is a resource with time lines of space missions, information about individual space vehicles and their missions, and biographies of astronauts.

The Planetary Society, www.planetary.org. The Planetary Society is the largest nonprofit, nongovernmental space advocacy group in the world. The society's Web site offers news archives, special features, current events, and information on how to get involved in space exploration activities.

SETI Institute, www.seti-inst.edu. SETI (Search for Extraterrestrial Intelligence) is the center for the study of life in the universe. Its Web site offers scientific papers, newsletters, and information on books written by SETI Institute scientists.

Space.com, www.space.com. Space.com offers special reports coverage of major space topics and events, as well as sections on spaceflight, science, technology, and SETI.

SpaceRef.com, www.spaceref.com. SpaceRef covers the history of astronomy and space exploration, including archives, essays, images, and press releases.

Students for the Exploration and Development of Space (SEDS), www.seds.com. SEDS is a student-based organization whose Web site includes space discussion forums, space image archives, astromaps, and a "Best of Hubble" section.

INDEX

airplanes, advances in, 145–46
Albee, Alden, 112
Aldrin, Buzz, 59, 150
American Society for Cell Biology
 (ASCB), 133
Anderson, Christopher, 111
Ansari, Rafat, 148
Apollo 11, 58, 63
Apollo Project, 16, 44
 descendants of, 63–64
 as political vs. scientific program,
 61
 products derived from, 145
 significance of, 58–64, 152
Armstrong, Neil, 17, 59, 80
Army Ballistic Missile Agency, 43
astroenvironmentalism, 156,
 157–59
Atlanta Constitution (newspaper),
 68
Atlantis (space shuttle), 83
Avdeyev, Sergei, 73, 76

Bahcall, John, 109
Bahcall, Neta, 108
Baldy (cartoonist), 68
Baraona, Cosmo, 148
Beach, Raymond, 147
Berghorn, Forrest, 71
Berkner, Lloyd, 35
Beutel, Allard, 107
Beyond Space Ship Earth:
 Environmental Ethics and the
 Solar System (Hargrove), 159
biomedicine, research in, 148–49
Blagonravov, Anatoli, 37
Bond, Peter, 46

Boynton, William, 112, 115
Briggs, Geoffrey, 120
Burrows, William E., 22, 158–59
Bush, George H.W., 70, 82, 141
Bush, George W., 82, 89
 calls for human exploration of
 space, 127

Campbell, Donald, 144, 145, 149
Cassini, 104
Challenger (space shuttle), 77–81
 crash of, implications for Mars
 missions, 120
 early problems of, 79
 mission of, 78
Chicago Tribune (newspaper), 68
Christensen, Philip, 112, 113, 116
Clarke, Arthur C., 14
Clementine series (military
 satellites), 116
Clinton, William, 139
 administration of, 140
Cohen, Jon, 111
Cold War, space race and, 15–16
Collins, Michael, 59
Columbia (space shuttle), 65–72
 crash of
 implications for Hubble Space
 Telescope, 107
 implications for ISS, 88–89
 lessons learned from, 82–89
 recovery of parts from, 84–85
commercial products
 from Apollo program, 145
 from space exploration, 15
communications satellites, 15, 44
 first, 16–17

computerized axial tomography (CAT) scanners, 15
Congress, U.S.
    funding for Mars missions and, 120–21
    suggestions for, on future of space exploration, 154–55
Conversation with the Starry Messenger (Kepler), 26
Copernicus, 23
costs
    of *Observer,* 113, 119–20
    reduction of, for space tourism, 153
    of space shuttles, 139
    for return-to-flight efforts, 86–87
Covey, Richard, 87
Crippen, Bob, 70–71
Crisp, Joy, 124
Cutter, W. Bowman, 19
Cyrano de Bergerac, Savinien, 25

Daily Mail (newspaper), 68
Deep Space Network, 121
DeLucas, Lawrence J., 133, 134
de Vries, Pete, 98
Die Rakete zu den Planetenraümen (Oberth), 30
Discovery (space shuttle), 17, 81
Dunning, John, 146

Edwin Hubble Telescope, 14
    accomplishments of, 108–109
Eisenhower, Dwight D., 36, 138
    creates NASA, 41
Endeavor (space shuttle), 146
energy storage technology, 147
Engle, Joe, 70
Explorer I (satellite), 40
Eyraud, Achille, 25

First Men on the Moon, The (Wells), 26
Flam, Faye, 106

Foale, C. Michael, 75, 89
Friedman, Herb, 37
From the Earth to the Moon (Verne), 26

Gagarin, Yuri, 46, 151
    background of, 47–48
    preparation for flight of, 48–49
    on view of Earth, 50
Galileo, 22, 24, 26
gallium arsenide semiconductor wafer, 141–42
Ganymede, *Voyager* exploration of, 102–103
Garber, Stephen J., 41
Garwin, James, 113
Gascoigne, William, 23
Gehman, Harold, 85, 89
Gemini Project, 41, 43–44
Ghosts of Mars, The (film), 158
Glenn, John H., Jr., 41, 44, 138, 144–45
Glenn Research Center, 144–49
global positioning networks, 15
Goddard, Robert H., 28, 29, 30–32
    achievements of, 32–33
Godwin, Francis, 25
Goldin, Daniel, 120
Gore, Al, 141
Greene, Nick, 77
Greenfield, Michael, 87
Grey, Jerry, 71–72
Grossman, Karl, 158–59

Hagan, William, 14
Halley, Edmund, 24
Halsell, Jim, 87
Hargrove, Eugene, 159
heliopause, 104, 105
Hinson, David, 113
Hoerter, Alicia, 67
Hoffman, Russell D., 19
Hubble, Edwin, 108
    see also Edwin Hubble Telescope

humans
  Mars settlement by, possibility of, 154
  microgravity is deleterious to health of, 134
  as space explorers are superior to robots, 126–30, 135–36, 141
  con, 135–36
Hutchinson, Neil, 71

intercontinental ballistic missiles (ICBMs), 35
International Geophysical Year (IGY), 34, 35, 42
International Space Station (ISS), 17
  experiments on, are inconsequential, 131–36
  implications of *Columbia* accident for, 88–89
  is scientifically unnecessary, 140–41
  maintenance of Hubble telescope and, 107
  as NASA jobs program, 139–40
Io, *Voyager* exploration of, 100–102
Isakeit, Diter, 17

James Webb Telescope, 14, 106, 109–10
Jarvis, Gregory B., 78
Jet Propulsion Laboratory (JPL), 43, 112
Johnson, Lyndon B., 38
Jupiter, *Voyager* exploration of moons of, 100–103
Jupiter-C (rocket), 40

Kaleri, Alexander, 89
Kamanin, Nikolai, 48
Kasyanov, Mikhail, 73, 76
Kennedy, John F., 54, 60, 127, 144, 151
  on spending for space program, 56–57
Kepler, Johannes, 26–27
Khrushchev, Nikita, 38
Klein, Michael, 118
Komarov, Vladimir, 53
Koptev, Yuri, 76
Korolev, Sergei, 35, 48
Kuznetz, Lawrence, 134

Land Ethic (Leopold), 159
Landsat satellites, 44
Lane, Lonne, 98
Launius, Roger D., 28, 41, 42
Le Figaro (newspaper), 69
Leopold, Aldo, 159
Lindbergh, Charles A., 32
Linenger, Jerry, 74–75
Liquid-Propellant Rocket Development (Goddard), 32
liquid rocket fuel, Goddard's work on, 30–32, 33
London, Josh, 137
London Times (newspaper), 69
Long, James, 98
Loukianos, 25
Ludd, Ned, 135

Magellan, 117
magnetic resonance imaging (MRI), 15
Magnuson, Ed, 65
Malin, Michael, 113
Man in the Moone, or a Discourse of a Voyager Thither by Domingo Gonzales, the Speedy Messenger, The (Godwin), 25
Mars
  future for missions to, 124
  life on, implications of, 92–93
  manned flights to, prospects of, 95–96
  mission to, public support for, 18
  as only destination for human exploration, 135

possibility of human settlement
    on, 154
rovers' exploration of, 122–24
water on
    evidence for, 123
    Sagan on, 93–94
Mars Environmental Survey
    (MESUR), 115
Mars Surveyor Lander mission, 147
Mather, John, 109–10
McAuliffe, Sharon Christa, 77, 78
McCormack, John, 39
McElroy, Neil, 39–40
McNair, Ronald E., 78
media, response of
    to *Columbia* flight, 68–69
    to Gargarin's flight, 52–53
    to *Sputnik I*, 38
    to *Vanguard* failure, 39
Mercury Project, 41, 43, 44
*Method of Reaching Extreme
    Altitudes, A* (Goddard), 31
microgravity
    is deleterious to human health, 134
    research in, 132–33
military programs, Apollo program
    supplants, 62–64
military satellites, 116
Miller, Ryder W., 156
Miner, Ellis, 98
Mir Space Station, 17, 73–76
    protein crystal experiments on,
        133
    records set by, 75–76
"Mission to Planet Earth" (MTPE),
    141
moon, the
    Kennedy vows to put man on,
        54–57
    robotic missions to, 44
    significance of, in human history,
        59–60
Morgan, Clay, 73
Morrison, David, 120

Morrison, Philip, 23

National Aeronautics and Space
    Administration (NASA), 14, 15
    criticism of, after *Challenger*
        disaster, 80
    formation of, 16, 41–44
    major programs of, 43–44
    Mars as goal for, 127–28
    projects of, are wasteful and
        inefficient, 137–42
        con, 143–49
National Research Council (NRC),
    133
Naval Research Laboratory, 43
Newton, Isaac, 26, 27
New York Herald Tribune
    (newspaper), 52
New York Times (newspaper), 31
Nixon, Richard M., 63

Oberth, Hermann, 28, 29–30
Observer
    cost of, 113, 119–20
    disappearance of, 111–21
    scientific experiments on, 114–15
O'Connell, Dennis, 67
O'Keefe, Sean, 82
O'Neill, Gerard, 67
Onizuka, Ellison S., 78
Opportunity, 122, 123–24, 128

Parazynski, Scott, 84
Park, Robert L., 13, 131
Pathfinder mission, 135
*Philosophiae Naturalis Principia
    Mathematica* (Newton), 27
Pickering, William, 37, 40
Pioneer 10, 102
Pioneer 11, 103
Polyakov, Valeri, 75–76
protein crystal experiments, 132–33

radiation belts, 34

Rayl, A.J.S., 97
R&D (magazine), 146
Readdy, Bill, 87
Reagan, Ronald, 80, 81
Red Mars, Green Mars, and Blue Mars (Robinson), 158
Red Planet (film), 158
Redstone (rocket), 40, 42
Resnick, Judith A., 78
Robinson, Kim Stanley, 158
robots
    limitations of, 128–29
    as space explorers, 18
        humans are superior to, 126–30, 135–36, 141
        con, 126–30
Rodgers, T.J., 142
Rogers, William, 80
Rudnev, Konstantin, 48

Sagan, Carl, 58, 94
    on implications of life on Mars, 92–93
    on organisms on Mars, 94–95
    on water on Mars, 93–94
satellites
    communications, 15, 44
    first, 16–17
    early programs for
        in U.S., 36
        in USSR, 35
    military, 116
Saturn, *Voyager* exploration of, 103–104
Schurmeier, Bud, 98
Science (journal), 113
Sheppard, Alan B., Jr., 41, 44, 55
Skylab, 44
Smith, Michael J., 78
Sojourner, 135
solar cells, Mars exploration and tests of, 147–48
Somnium (Kepler), 26
Soviet Union

launches first man in orbit, 46–53
satellite program of, 35
see also Mir Space Station
Soyuz (spacecraft), 53
space
    exploration of
        controversy over, 18–19
        early fantasies of, 25–26
        precursors to, 22–27
        robots should be used for, 126–30
            con, 126–30, 135–36, 141
        should be supported by space tourism, 150–55
        technical and scientific value of, 14–15
    first human in, 46–53
    junk, 19, 157
    as next war zone, 159
    should be protected as a wilderness, 156–60
space shuttles, 44, 78, 138
    costs of, 86–87, 139
    see also names of individual crafts
spacewalk, first, 44
Spirit, 122, 123, 128
Sputnik I, 4, 34, 42–43
    launch of, 37–38
    public reaction to, 43
Squyres, Steve, 122
Stafford, Thomas, 87
Starcraft Boosters, 150
Starry Messenger, The (Galileo), 22, 24, 26
Stembridge, Charles, 98
Stone, Ed, 97
Sullivan, Walter, 36–37
Sun (newspaper), 68
surveys, on support for Mars mission, 18
Suzuki, Zenko, 69
Swift, Jonathan, 25

Takhtarova, Anya, 47

Teacher in Space Program (TISP), 48

technology

advances in, from space program, 145

in energy storage, 147

Apollo program inspired optimism about, 64

telescopes

of Galileo, 23

see also Edwin Hubble Telescope; James Webb Telescope

Telstar, 16

Thagard, Norm, 75

This New Ocean (Burrows), 158

Titov, Vladimir, 49, 76

tourism, space, 150–55

Travis, John, 111

True History (Loukianos), 25

Tsiolkovsky, Konstantin Eduardovich, 27, 28, 29

UNISPACE III (Third United Nations Conference on the Exploration and Peaceful Uses of Outer Space), 160

United States, early satellite program of, 36

Uranus, Voyager exploration of, 104

U.S. News & World Report (magazine), 91

USSR. See Soviet Union

Van Allen, James, 35, 37, 40

Van Allen Radiation Belts, discovery of, 40, 43

Vanguard Project, 42

Vanguard (rocket), 34, 36, 42

launch of, 39–40

Venus

Magellan mission to, 117

robotic missions to, 44

Verne, Jules, 22, 26, 31

Viking missions, 91–96, 129

von Braun, Wernher, 30, 39–40

Voyager 1, 97

data from, 99–100

Voyager 2, 97, 102

Voyages to the Moon and the Sun (Cyrano de Bergerac), 25

Voyage to Venus (Eyraud), 25

Walter, William J., 34

Walton, Marsha, 122

Wells, H.G., 22, 26

Wheelon, Albert, 139

White, Edward H., Jr., 41, 44

Wiesner, James, 61

Wilkinson, Todd, 143

Williamson, Ray A., 28

Wolf, David, 75

Wolfe, Tom, 67

Wrong Stuff, The (Grossman), 158

Yeager, Chuck, 80

Young, John, 65, 68, 70–71

Youngquist, Wayne, 68

Zubrin, Robert, 126

Zvezdniy Gorodok (Soviet training center), 48